# Modern Organic Synthesis in the Laboratory

# Modern Organic Synthesis in the Laboratory

*A Collection of Standard Experimental Procedures*

Jie Jack Li
Chris Limberakis
Derek A. Pflum
*Pfizer Global Research & Development*

# OXFORD
UNIVERSITY PRESS

2007

# OXFORD
UNIVERSITY PRESS

Oxford University Press, Inc., publishes works that further
Oxford University's objective of excellence
in research, scholarship, and education.

Oxford New York
Auckland Cape Town Dar es Salaam Hong Kong Karachi
Kuala Lumpur Madrid Melbourne Mexico City Nairobi
New Delhi Shanghai Taipei Toronto

With offices in
Argentina Austria Brazil Chile Czech Republic France Greece
Guatemala Hungary Italy Japan Poland Portugal Singapore
South Korea Switzerland Thailand Turkey Ukraine Vietnam

Copyright © 2007 by Oxford University Press, Inc.

Published by Oxford University Press, Inc.
198 Madison Avenue, New York, New York 10016

www.oup.com

Oxford is a registered trademark of Oxford University Press

All rights reserved. No part of this publication may be reproduced,
stored in a retrieval system, or transmitted, in any form or by any means,
electronic, mechanical, photocopying, recording, or otherwise,
without the prior permission of Oxford University Press.

Library of Congress Cataloging-in-Publication Data
Li, Jie Jack.
Modern organic synthesis in the laboratory: a collection of standard
experimental procedures / Jie Jack Li, Chris Limberakis, and Derek A. Pflum.
  p. cm.
Includes bibliographical references and index.
ISBN 978-0-19-518798-4; 978-0-19-518799-1 (pbk.)
1. Organic compounds—Synthesis. I. Limberakis, Chris.
II. Pflum, Derek A. III. Title.
QD262.L47 2007
547'.2078—dc22  2006036964

The material in this text is intended to provide general guidelines and is for informational
purposes only. Although the materials, safety information, and procedures described
in this book are designed to offer current, accurate and authoritative information
with respect to the subject matter covered, the information and these procedures
should serve only as a starting point for laboratory practices. They do not purport to
specify minimal legal standards or to represent the policy of the authors or Oxford
University Press.

This book is sold with the understanding that no warranty, guarantee, or representation
is made by the authors or Oxford University Press as to accuracy or specificity of the
contents of this material and that they make no warranties of merchantability or
fitness for a particular purpose. The authors and Oxford University Press specifically
disclaim any responsibility for any loss, injury, or damages that may be claimed or
sustained by anyone as a consequence, directly or indirectly, of the use and/or application
of any of the contents in this book.

9 8 7 6 5 4 3 2 1

Printed in the United States of America
on acid-free paper

*To Sherry, Rachele, and Mary Kay*
*For Their Love, Support, and Patience*

# Foreword

*Those who do not want to imitate anything, produce nothing.*
—Salvador Dali

There exists a close analogy between the trials and tribulations of a chemist and those of an artisan. Just as an artist must balance the sensitive interplay of color, light, and brushstroke to create a timeless painting, so too must the chemist carefully marshal the reagents, solvents, and reactions that occupy chemical space to successfully forge a new molecular entity. Finding reliable, standardized procedures for those routine transformations is often a time-consuming and tedious process, one that subtracts from effort that could be devoted towards more creative and exciting pursuits. Indeed, much of the excitement, fascination, and allure of synthetic organic chemistry resides in the imaginative exploration of new realms in chemical reactivity and complexity, not the search for standard operation procedures.

In *Modern Organic Synthesis in the Laboratory*, Drs. Jie Jack Li, Chris Limberakis, and Derek Pflum provide procedures for the common transformations and protocols that permeate the tapestry of synthetic organic chemistry. From guidelines of proper flash chromatography technique, to procedures for drying solvents, and recipes for staining solutions, all of the basic, yet essential, organic laboratory techniques are described in detail. Nearly all types of common reactions for oxidations, reductions, functional group manipulations, and C–C bond formation are covered. With little question, this book is beautifully suited for practitioners of the field at all levels of study because of the thorough selection of procedures and a layout that is both well designed and comprehensive.

We congratulate the authors for filling such an important gap in the chemical literature by carefully designing what will undoubtedly become a survival manual destined for untold numbers of organic chemists.

K. C. Nicolaou and Phil S. Baran<br>
The Scripps Research Institute<br>
July 12, 2006

# Preface

The prevalence of computer databases has, at the present time, made literature searching in organic synthesis an increasingly easier enterprise. However, the dilemma has now become how to select one procedure in an ocean of choices. For novices in the laboratory, in particular, this is a daunting task, deciding what reaction conditions to try first in order to have the best chance of success. This collection intends to serve as an "older and wiser" lab-mate by sharing the authors' own experience through this compilation of the most commonly used experimental procedures from established groups and/or rigorously reviewed journals.

Under the title of each experimental procedure, brief commentaries are often offered which summarize the authors' personal experience, and in many instances review articles are cited. These experimental procedures and commentaries are by no means "carved in stone", as you and your lab-mates may have your own favorite protocol. For the final products, detailed spectral data are not given because they simply take up too much space.

We are indebted to Professor John P. Wolfe at the University of Michigan, and his students, Joshua Ney and Josephine Nakhla; Professor Phil S. Baran at the Scripps Research Institute, and his students, Noah Z. Burns, Mike DeMartino, Tom Maimone, Dan O'Malley, Jeremy Richter, and Ryan Shenvi for proofreading the manuscript.

We had a good time putting together these experimental procedures. We have even been using the manuscript ourselves quite often. Hopefully you will find it as useful. We welcome your critique!

Jack Li, Chris Limberakis, and Derek Pflum
Ann Arbor, Michigan
September 4, 2006

# Acknowledgments

The authors would like to extend their sincerest gratitude to the original publishers of the experimental procedures for granting permission to reprint these materials that form the core of this book. The publishers include: The American Chemical Society (*The Journal of the American Chemical Society, Journal of Medicinal Chemistry, Journal of Organic Chemistry, Organic Letters, Organic Process Research and Development*, and *Organometallics*), Elsevier (*Biorganic Medicinal Chemistry, Journal of Organometallic Chemistry, Tetrahedron*, and *Tetrahedron Asymmetry*), John Wiley & Sons (*Organic Syntheses*), Royal Society of Chemistry (*Journal of the Chemical Society Perkin Transactions 1*), Wiley-VCH Verlag GmbH & Co KG (*Chemistry-A European Journal*), Taylor & Francis Group (*Synthetic Communications*), NRC Research Press (*Canadian Journal of Chemistry*), Thieme (*Synlett*), and The Japan Institute of Heterocyclic Chemistry (*Heterocycles*). In addition, we would like to express much appreciation to the following companies for allowing us to use their product images in Chapter 1: Innovative Technology (Pure-Solv 400 Solvent Purification System), Analogix (Flash 12/40 system$^{TM}$), Teledyne Isco (CombiFlash®Companion®), and Biotage (Flash 400$^{TM}$).

# Contents

Foreword  vii

Abbreviations and Acronyms  xv

**1**  Fundamental Techniques  3
    1.1  Safety!  3
    1.2  Useful Preparations  5
    1.3  Chromatography  17
    1.4  Crystallization  22
    1.5  Residual Solvent Peaks in Nuclear Magnetic Resonance  24

**2**  Functional Group Manipulations  28
    2.1  Alcohol Oxidation State  28
    2.2  Ketone Oxidation State  39
    2.3  Acid Oxidation State  44

**3**  Oxidation  55
    3.1  Alcohol to Ketone Oxidation State  55
    3.2  Alcohol to Acid Oxidation State  68
    3.3  Olefin to Diol  70
    3.4  Aldehyde to Acid Oxidation State  76
    3.5  Heteroatom Oxidations  78

**4　Reductions　81**

    4.1　Alcohols to Alkanes　81
    4.2　Aldehydes, Amides, and Nitriles to Amines　83
    4.3　Carboxylic Acids and Derivatives to Alcohols　85
    4.4　Esters and Other Carboxylic Acid Derivatives to Aldehydes　88
    4.5　Ketones or Aldehydes to Alcohols　91
    4.6　Ketones to Alkanes or Alkenes　98
    4.7　Reductive Dehalogenations　104
    4.8　Carbon–Carbon Double and Triple Bond Reductions　105
    4.9　Heteroatom–Heteroatom Reductions　108

**5　Carbon–Carbon Bond Formation　111**

    5.1　Carbon–Carbon Forming Reactions (Single Bonds)　111
    5.2　Carbon–Carbon Double Bonds (Olefin Formation)　150
    5.3　Reactions that Form Multiple Carbon–Carbon Bonds　166

**6　Protecting Groups　168**

    6.1　Alcohols and Phenols　168
    6.2　Amines and Anilines　176
    6.3　Aldehydes and Ketones　185

    Index　191

# Abbreviations and Acronyms

| | | | |
|---|---|---|---|
| ◯— | polymer support | BINALH | Lithium 2,2´-dihydroxy-1,1´-binaphthylethoxy-aluminum hydride |
| Ac | acetyl | | |
| acac | acetylacetonate | | |
| AcOH | acetic acid | BINAP | 2,2-bis(diphenylphosphino)-1,1´-binaphthyl |
| AE | asymmetric epoxidation reaction | | |
| | | BINOL | 1,1´-bi-2-naphthol |
| AIBN | 2,2´-azobisisobutyronitrile | BMS | borane dimethyl sulfide complex |
| Alpine-borane® | B-isopinocamphenyl-9-borabicyclo[3.3.1]-nonane | Bn | benzyl |
| | | Boc | tert-butyloxycarbonyl |
| | | BOM | benzyloxymethyl |
| AME | acetyl malonic ester | BOP | benzotriazol-1-yloxy-tris(dimethylamino)-phosphonium hexafluorophosphate |
| AMNT | aminomalononitrile p-toluenesulfonate | | |
| Ar | aryl | | |
| B: | generic base | BPO | benzoyl peroxide |
| 9-BBN | 9-borabicyclo-[3.3.1]nonane | Bu | butyl |
| | | t-Bu | tert-butyl |
| BFO | benzofurazan oxide | Bz | benzoyl |
| TBHP | tert-butyl hydrogen peroxide | °C | degree Celsius |
| | | CAN | ceric ammonium nitrate (ammonium cerium(IV) nitrate) |
| BHT | butylated hydroxy toluene | | |

## ABBREVIATIONS AND ACRONYMS

| | | | |
|---|---|---|---|
| Chirald® | (2S,3R)-(+)-4-dimethylamino-1,2-diphenyl-3-methyl-2-butanol | (DHQD)₂-PHAL | 1,4-bis(9-O-dihydroquinidine)-phthalazine |
| CTAB | cetyl trimethylammonium bromide | DIAD | diisopropyl azodicarboxylate |
| CBS | Corey–Bakshi–Shibata | DIBAL-H | diisobutylaluminum hydride |
| Cbz | benzyloxycarbonyl | DIC | diisopropylcarbodiimide |
| cp | cyclopentadienyl | Diglyme | diethylene glycol dimethyl ether |
| CSA | camphorsulfonic acid | | |
| CuTC | copper thiophene-2-carboxylate | Dimsyl | methylsulfinylmethide |
| cy | cyclohexyl | DIPEA | diisopropylethylamine |
| DABCO | 1,4-diazabicyclo[2.2.2]octane | DMAc | N,N-dimethylacetamide |
| DAPA | dipotassium azodicarboxylate | DMA | N,N-dimethylaniline |
| DAST | (diethylamino)sulfur trifluoride | DMAP | N,N-dimethylaminopyridine |
| dba | dibenzylideneacetone | DMDO | dimethyldioxirane |
| DBE | 1,2-dibromoethane | DME | 1,2-dimethoxyethane |
| DBU | 1,8-diazabicyclo[5.4.0]undec-7-ene | DMF | dimethylformamide |
| DBN | 1,5-diazabicyclo[4.3.0]non-5-ene | DMFDMA | dimethylaminoformaldehyde dimethyl acetal |
| DCB | dichlorobenzene | | |
| DCC | 1,3-dicyclohexylcarbodiimide | DMP | Dess-Martin periodinane |
| DCM | dichloromethane | DMPU | N,N'-dimethyl-N,N'-propylene urea |
| DDQ | 2,3-dichloro-5,6-dicyano-1,4-benzoquinone | DMS | dimethylsulfide |
| | | DMSO | dimethylsulfoxide |
| de | diastereomeric excess | DMSY | dimethylsulfoxonium methylide |
| DEAD | diethyl azodicarboxylate | DMT | dimethoxytrityl |
| DEPC | diethyl phosphorocyanidate | DNP | 2,4-dinitrophenyl |
| | | L-DOPA | 3,4-dihydroxyphenylalanine |
| DET | diethyl tartrate | | |
| Δ | solvent heated under reflux | DPPA | diphenylphosphoryl azide |
| DHP | dihydropyran | dppb | 1,4-bis(diphenylphosphino)butane |
| DHPM | 3,4-dihydropyrimidin-2(1H)-one | | |
| (DHQ)₂-PHAL | 1,4-bis(9-O-dihydroquinine)-phthalazine | dppe | 1,2-bis(diphenylphosphino)ethane |

| | | | |
|---|---|---|---|
| dppf | 1,1′-bis(diphenylphosphino)ferrocene | GC | gas chromatography |
| | | glyme | 1,2-dimethoxyethane |
| dppp | 1,3-bis(diphenylphosphino)propane | HOBt | 1-hydroxybenzotriazole |
| | | h | hour(s) |
| dr | diastereomeric ratio | hv | irradiation with light |
| | | His | histidine |
| $E$ | Entgegen (opposite, trans) | HMDS | hexamethyldisilazine |
| | | HMPA | hexamethylphosphoramide |
| E1 | unimolecular elimination | HMPT | hexamethylphosphorous triamide |
| E2 | bimolecular elimination | HOMO | highest occupied molecular orbital |
| E1cb | 2-step, base-induced β-elimination via carbanion | HPLC | high performance liquid chromatography |
| EDA | ethylendiamine | IBCF | isobutylchloroformate |
| EDCI | 1-ethyl-3-[3-(dimethylamino)propyl]carbodiimide hydrochloride | IBX | 1-Hydroxy-1,2-benziodoxol-3(1$H$)-one |
| | | Imd | imidazole |
| EDG | electron donating group | IPA | isopropanol |
| EDTA | ethylenediamine tetraacetic acid | $i$-Pr | isopropyl |
| | | KHMDS | potassium hexamethyldisilazide |
| $ee$ | enantiomeric excess | kg | kilogram(s) |
| EEDQ | 2-ethoxy-1-ethoxycarbonyl-1,2-dihydroquinoline | K-selectride® | potassium tri-$sec$-butylborohydride |
| | | L | liter(s) |
| EMME | ethoxymethylenemalonate | LAH | lithium aluminum hydride |
| ent | enantiomer | LDA | lithium diisopropylamide |
| EPP | ethyl polyphosphate | | |
| eq | equivalent | LHMDS | lithium hexamethyldisilazide |
| Et | ethyl | | |
| EtOAc | ethyl acetate | LiHMDS | lithium hexamethyldisilazide |
| EPR (= ESR) | electron paramagnetic resonance spectroscopy | L-selectride® | lithium tri-$sec$-butylborohydride |
| ESR (= EPR) | electronic spin resonance | LTMP | lithium 2,2,6,6-tetramethylpiperidine |
| EWG | electron withdrawing group | LUMO | lowest unoccupied molecular orbital |
| FMO | frontier molecular orbital | M | metal |
| Fmoc | 9-fluorenylmethoxycarbonyl | M | moles per liter (molar) |
| | | MCR | multi-component reaction |
| FVP | flash vacuum pyrolysis | | |
| g | gram(s) | $m$-CPBA | $m$-chloroperoxybenzoic acid |
| GABA | γ-aminobutyric acid | | |

| | | | |
|---|---|---|---|
| Me | methyl | PPA | polyphosphoric acid |
| MEM | β-methoxyethoxymethyl | PPE | personal protection equipment |
| Mes | mesitylenyl | | |
| MET | methyl ethyl ketone | PPE | polyphosphoric ester |
| μg | microgram(s) | 4-PPNO | 4-phenylpyridine-$N$-oxide |
| μL | microliter(s) | PPP | 3-(3-hydroxyphenyl)-1-$n$-propylpiperidine |
| μmol | micromole(s) | | |
| mg | milligram(s) | PPSE | polyphosphoric acid trimethylsilyl ester |
| mL | milliliter(s) | | |
| mmol | millimole(s) | PPTS | pyridinium $p$-toluenesulfonate |
| MMPP | magnesium monoperoxyphthalate hexahydrate | | |
| | | Pr | propyl |
| MO | molecular orbital | Pro | proline |
| mol | mole(s) | psi | pounds per square inch |
| MOM | methoxymethyl | PTC | phase transfer catalyst |
| Ms | methanesulfonyl (mesyl) | $p$-TSA | $para$-toluenesulfonic acid |
| MS | molecular sieves | | |
| MSDS | material safety data sheet | Py or Pyr | pyridine |
| MTBE | methyl $tert$-butyl ether | Ra-Ni | Raney nickel |
| MTPA | α-methoxy-α-trifluoromethylphenylacetic acid | RCM | ring-closing metathesis |
| | | Redal-H | sodium bis(2-methoxyethoxy(aluminum hydride) |
| MVK | methyl vinyl ketone | | |
| MWI (μν) | microwave irradiation | | |
| NBS | $N$-bromosuccinimide | ROM | ring-opening metathesis |
| NCS | $N$-chlorosuccinimide | rt | room temperature |
| NIS | $N$-iodosuccinimide | Salen | $N,N'$-disalicylideneethylenediamine |
| NMDA | $N$-methyl-D-aspartate | | |
| NMM | $N$-methylmorpholine | SEM | 2-(trimethylsilyl)ethoxymethyl |
| NMO | $N$-methylmorpholine-$N$-oxide | | |
| | | SET | single electron transfer |
| NMP | 1-methyl-2-pyrrolidinone | | |
| NMR | nuclear magnetic resonance | $S_NAr$ | nucleophilic aromatic substitution |
| | | $S_N1$ | unimolecular nucleophilic substitution |
| Nu | nucleophile | | |
| PCC | pyridinium chlorochromate | $S_N2$ | bimolecular nucleophilic substitution |
| PDC | pyridinium dichromate | | |
| PDE | phosphodiesterase | $t$-Bu | $tert$-butyl |
| PEG | polyethylene glycol | TADDOL | α,α,α´,α´-tetraaryl-4,5-dimethoxy-1,3-dioxalane |
| pGlu | pyroglutamic acid | | |
| Ph | phenyl | | |
| PhFl | 9-phenylfluoren-9-yl | TASF | $(Et_2N)_3S^+(Me_3SiF_2)^-$ |
| phth | phthaloyl | TBAF | tetrabutylammonium fluoride |
| pKa | log acidity constant | | |
| PMA | phosphomolybdic acid | TBD | 1,5,7-triazabicyclo[4.4.0]dec-5-ene |
| PMB | $para$-methoxybenzyl | | |

| | | | |
|---|---|---|---|
| TBDMS or TBS | *tert*-butyldimethylsilyl | THP | tetrahydropyranyl |
| TBDPS | *tert*-butyldiphenylsilyl | TIPS | triisopropylsilyl |
| | | TLC | thin layer chromatography |
| TBHP | *tert*-butylhydroperoxide | TMEDA | $N,N,N',N'$-tetramethylethylenediamine |
| TCCA | trichlorocyanuric acid | TMG | tetramethylguanidine |
| | | TMP | tetramethylpiperidine |
| TCT | 2,4,6-trichloro-[1,3,5]-triazine | TMS | trimethylsilyl |
| | | TMSCl | trimethylsilyl chloride |
| TEA | triethylamine | | |
| TEMPO | 2,2,6,6-tetramethyl-1-piperidinyloxy, free radical | TMSCN | trimethylsilyl cyanide |
| | | TMSI | trimethylsilyl iodide |
| TES | triethylsilyl | TMSOTf | trimethylsilyl triflate |
| Tf | trifluoromethanesulfonyl (triflic) | Tol | toluene or tolyl |
| | | Tol-BINAP | 2,2´-bis(di-*p*-tolylphosphino)-1,1´-binaphthyl |
| TFA | trifluoroacetic acid | | |
| TFAA | trifluoroacetic anhydride | TosMIC | (*p*-tolylsulfonyl)methyl isocyanide |
| TFE | trifluoroethanol | TPAP | tetra-*n*-propylammonium perruthenate |
| TfOH | triflic acid | | |
| TFP | tri-*o*-furylphosphine | Tr | trityl |
| TFPAA | trifluoro peracetic acid | TRIS | tris(hydroxymethyl)aminomethane |
| TFSA | trifluorosulfonic acid | Ts(Tos) | *p*-toluenesulfonyl (tosyl) |
| THF | tetrahydrofuran | TSA | *p*-toluenesulfonic acid |
| THP | tetrahydropyran | TsO | tosylate |
| THIP | 4,5,6,7-tetrahydroisoxazolo[5,4-c]-pyridin-3-ol | $X_c$ | Chiral auxillary |
| | | Z | Zusammen (together, cis) |

# Modern Organic Synthesis in the Laboratory

# 1

# Fundamental Techniques

## 1.1 Safety!

What we do in a modern organic chemistry laboratory is serious business. While it can provide social benefit, basic scientific discoveries, and intellectual satisfaction, chemical experiment is not just fun, it can also be very hazardous, some experiments inherently so. Complacency is often observed by veterans and novices alike. One often forgets that chemistry is a potentially dangerous enterprise; a cavalier attitude often results in disastrous consequences. Therefore, extreme caution should be exercised at all time, especially when one handles large-scale reactions that are exothermic or when dealing with toxic chemicals.

### 1.1.1 Personal Protection Equipment

#### 1.1.1.1 Safety Glasses

If a chemical splashes into your eyes, it could do serious and sometimes permanent damage to your vision. The most common forms of eye protection include safety glasses (with sideshields), goggles, and face shields. Prescription eye glasses are acceptable provided that the lenses are impact resistant and they are equipped with side shields.

While at the Massachusetts Institute of Technology, Professor K. Barry Sharpless, the 2001 chemistry Nobel laureate, experienced an event that forever changed his life. Professor Sharpless normally wore his safety glasses, but one evening in 1970 he was examining a sealed nuclear magnetic resonance (NMR) tube without safety glasses. Unfortunately for Professor Sharpless, the tube exploded, spraying glass fragments into one of his eyes. The damage was so severe that he lost functional vision in the injured eye. Professor Sharpless's own words summarize the importance of eye protection, "The lesson to be learned from my experience is straightforward: there's simply never an adequate excuse for not wearing safety glasses in the laboratory at all times" (Scripps Research Institutes' Environmental Health and Safety Department Safety Gram, 2000 (2nd quarter), www.scripps.edu/researchservices/ehs/News/safetygram/).

### 1.1.1.2 Gloves

Laboratory gloves are an essential part of safe laboratory practice and *must* be worn while handling chemicals. Despite practicing good safety techniques, tragedy may still strike. In 1997, Dr. Karen Wetterhahn, a world-renowned Dartmouth College chemistry professor, died of mercury poisoning 10 months after as little as a drop (0.1 mL) of dimethylmercury [$Hg(CH_3)_2$] seeped through her latex gloves [(a) Blayney, M. B.; Winn, J. S.; Nierenberg, D. W. *Chem. Eng. News* **1997**, *75(19)*, 7. (b) Nagal, M. C. *Chem Health Safety* **1997**, *4*, 14–18]. Although she was working in a ventilated hood and was gowned with personal protection equipment (PPE), independent tests showed that her latex gloves provided virtually no protection against dimethylmercury. This tragic event illustrates that gloves are not a perfect barrier to chemicals, care must still be taken to minimize exposure, and the *proper* gloves must be worn. When choosing the type of glove, the researcher should consider several factors, including degradation and permeation by the chemical, type of exposure, temperature, glove thickness, and physical resistance of the glove. Table 1.1 summarizes the types of gloves available and their recommended use. The manufacturer's description of the gloves should be consulted. In addition, glove selection references include (a) Mellstrom, G. A.; Wahlberg, J. E.; Maibach, H. I. *Protective Gloves for Occupational Use*; CRC: Boca Raton, FL, 1994. (b) http://www.bestglove.com (accessed 30 Jan 2007).

### 1.1.1.3 Laboratory Coats

Laboratory coats provide an important barrier for your clothes and, more important, your skin from chemicals. The laboratory coat should fit comfortably, have long sleeves, and should be clean.

Table 1.1 Types of gloves and recommended uses

| Glove type | Recommended |
|---|---|
| Latex | Dilute acids and bases |
| Butyl | Acetone, $CH_3CN$, DMF, DMSO |
| Neoprene | Acids, bases, peroxides, hydrocarbons, alcohols, phenols |
| Nitrile | Acetic acid, acetonitrile, DMSO, ethanol, ether, hexane, dilute acids |
| Polyvinyl alcohol | Aromatic and chlorinated solvents |
| Polyvinyl chloride | Acids, bases, amines, peroxides |
| Viton | Chlorinated solvents, aromatic solvents |
| Silver shield | Wide variety of chemicals, provides the highest level of protection |

DMF, dimethylformamide; DMSO, dimethylsulfoxide.

## 1.1.2 Material Safety Data Sheets

When dealing with chemicals, caution is warranted, especially with reactive chemicals, carcinogens, and toxic reagents. A useful resource is the Material Safety Data Sheets (MSDS). The MSDS can be considered the "identification card" for a given reagent. A researcher should exercise due diligence and read the MSDS to have a better understanding of the chemical. An MSDS provides a plethora of data, including chemical and physical properties, health hazards, first aid measures in case of exposure, fire fighting measures, handling/storage, stability/reactivity, recommended PPE, toxicological data, and other pertinent information specific to the given chemical. Nowadays, many chemicals come with MSDS sheets, enabling us to scrutinize them before use in our reactions. There are a variety of resources that can be accessed online, including the following: www.ilpi.com/msds (accessed Jan. 30, 2007), MSDS Solution [www.msds.com (accessed Jan. 30, 2007)], MSDS online [www.msdsonline.com (accessed Jan. 30, 2007)], Seton Compliance Research Center [www.setonresourcecenter.com (accessed Jan. 30, 2007)], Cornell University [msds.ehs.cornell.edu (accessed Jan. 30, 2007)], Vermont SIRI [hazard.com (accessed Jan. 30, 2007)], as well as chemical suppliers/manufactures such as Sigma-Aldrich (www.sigmaaldrich.com), VWR (www.vwrsp.com), and others.

Gone are the days when a chemist could smoke a cigarette in the laboratory. Arthur J. Birch was photographed smoking a cigar while demonstrating an ether extraction, which is unthinkable today.

## 1.1.3 Never Taste Chemicals

In the "good old days," chemists routinely tasted newly synthesized compounds and documented their taste as part of the scientific record. This practice may explain why so many prominent chemists suffered poor health during that time. In the 19th century, the German chemist Justus von Liebig once declared, "A chemist with good health must not be a good chemist" (Li, J. J. *Laughing Gas, Viagra, and Lipitor: The Human Stories behind the Drugs We Use*; Oxford University Press: New York; 2006, p. 79). In 1965, a chemist at G. D. Searle, Jim Schlatter, accidentally had a small amount of a compound on his hands without noticing it. Later that morning, he licked his finger as he reached for a piece of paper and noticed a sweet taste. After careful analysis, he discovered that the compound was the methyl ester of the dipeptide of L-aspartic acid and L-phenylalanine that was later marketed as the sugar substitute Nutrasweet® (aspartame), a multibillion dollar a year product. Schlatter was lucky. If the by-product had been an extremely toxic chemical, we would have lost a chemist rather than gaining a food additive.

## 1.2 Useful Preparations

Setting up the reaction is probably the most import job for an organic chemist. Once a reaction is initiated, there is little left that needs to be done to change the outcome. Individual reaction conditions are surveyed in the ensuing chapters. Herein, anhydrous solvents and a list of useful cooling baths for maintaining reaction temperatures below 0 °C are provided. In addition, several important preparations of commonly used organic reagents, including Grignard reagents, organolithium reagents, organozinc reagents, diazomethane, Dess-Martin reagent, preparation of lithium diisopropylamide (LDA), and Jones reagent, are described.

### 1.2.1 Anhydrous Solvents

Traditionally, anhydrous solvents are obtained from distilling the solvent from a drying agent under an inert atmosphere (i.e., nitrogen or argon). The distilled solvent is then collected in a reservoir and removed via a syringe or transferred directly to a round-bottomed flask (Figure 1.1). Table 1.2 lists typical solvents and the drying agents.

Also, there is a misconception concerning solvents distilled from sodium/benzophenone. For example, tetrahydrofuran (THF) is distilled from sodium/benzophenone. Prior to obtaining "dry" THF, the solvent in the distillation pot is a characteristic blue or sometimes purple in color because of the thus afforded benzophenone ketyl radical. It has been assumed by many that the blue color is indicative of "water free" THF. In fact, this may not be the case. Mallinkrodt-Baker conducted a study that showed that the blue color was an indication of the absence

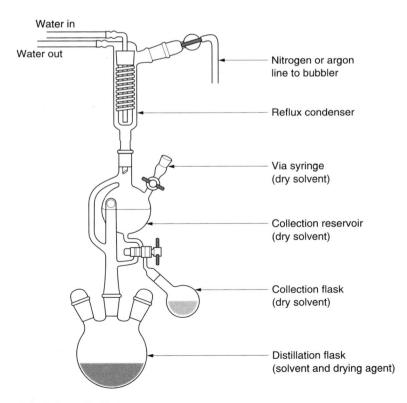

Figure 1.1 Solvent distillation setup.

## FUNDAMENTAL TECHNIQUES  7

Table 1.2 Anhydrous solvents from distillation

| Solvent | Drying agent |
|---|---|
| Dichloromethane | Calcium hydride |
| Diethyl ether | Sodium/benzophenone |
| Tetrahydrofuran | Sodium/benzophenone |
| Acetonitrile | Calcium hydride |
| Ethanol or methanol | Magnesium/iodine |
| Toluene | Calcium hydride or sodium |
| Benzene | Calcium hydride or sodium |
| Triethylamine | Calcium hydride |

of dissolved oxygen, not water (see "Benzophenonone Ketyl Study," www.mallbaker.com/techlib/). It required approximately 8 mol% oxygen to quench the benzophenone radical color; however, a full 100% mol of water was added before the blue color dissipated. In addition, the breakdown of the benzophenone radical delivered benzene as the major impurity (150 ppm), as well as other impurities, including triphenylmethanol, phenol, and diphenylmethane. Any one of these impurities could adversely affect a given reaction.

Although distillation is an effective drying technique, it is laden with serious safety issues. They include fire hazards, the handling of large quantities of reactive metals or metal hydrides, quenching of the metals or metal hydrides, and waste disposal.

Instead of using solvent stills there are other options. Perhaps the least expensive method is to dry the solvent over activated 4 Å molecular sieves. This is a particularly effective method when small quantities of solvent are required (<100 mL). Also, there are now commercial sources of a variety of anhydrous solvents. Some solvents include acetonitrile, chloroform, dichloromethane, dimethylsulfoxide (DMSO), 1,4-dioxane, dimethylformamide (DMF), diethyl ether, hexane, ethyl acetate, ethanol, isopropanol, methanol, 1-methyl-2-pyrrolidinone, 2-methylTHF, THF, and toluene. Although there is a cost disadvantage, these solvents are usually suitable for most reactions requiring an anhydrous solvent. Finally, a method that many institutions have adapted is solvent purification systems. This method was popularized after the ground-breaking paper by Grubbs and coworkers in 1996 (Pangborn, A. B.; Giardello, M. A.; Grubbs, R. H.; Rosen, R. K.; Timmers, F. J. *Organometallics* **1996**, *15*, 1518–1520). Although homemade units can be constructed, commercial instruments are available (Figure 1.2). In a typical set up, ultra dry solvent in a flammable proof solvent cabinet is pushed through two columns, typically alumina and copper catalyst columns, with nitrogen or argon. The anhydrous and oxygen free solvent is now ready for removal via a flask or syringe. These are far safer than solvent stills; however, they too have disadvantages, including the initial cost and a pressurized system.

In the future, anhydrous solvents will more than likely be obtained from all of the methods described but one should always choose wisely. If a safer alternative is available, one should opt for it.

In the past 15 years, the organic chemistry community has been very cognizant of the environmental impact of organic solvents and thus green solvents are growing in importance. Green solvents are defined as solvents that have minimal toxicity to

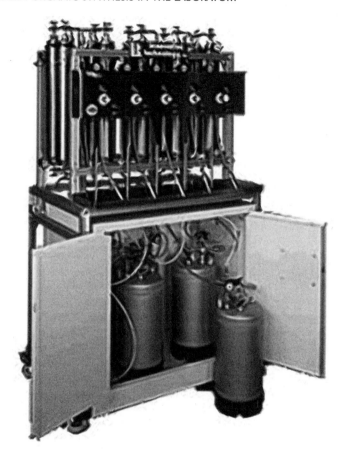

Figure 1.2 Commercial solvent purification unit: Pure-Solv 400 Solvent Purification System (reprinted with permission from Innovative Technology).

humans and the environment where their toxicities are well understood (Nelson, W. M. *Green Solvents for Chemistry Perspectives and Practice*; Oxford University Press: Oxford; 2003, pp. 91–92). Green solvents that are growing in importance include ionic liquids, fluorous solvents, supercritical carbon dioxide, water, ethanol, and aqueous miscelles and polymers (Mikami, K., ed. *Green Reaction Media in Organic Synthesis*: Blackwell: Oxford, 2005). Although green solvents are nontraditional, there are traditional solvents that are considered green, including acetic acid, benzyl benzoate, diethyl glycol dimethyl ether, DMSO, ethyl acetate, glycerol, hexane, methanol, *t*-butanol, and THF (Nelson, W. M. *Green Solvents for Chemistry Perspectives and Practice*; Oxford University Press: Oxford; 2003, p. 213). In addition to green solvents, organic chemists have identified alternatives (or drop-in) solvents which are considered safer alternatives for a variety of reasons. Table 1.3 illustrates some of these alternatives.

Table 1.3 Alternative solvents

| Traditional solvent | Issue with solvent | Drop-in solvent |
|---|---|---|
| Benzene | Carcinogen | Toluene |
| Carbon tetrachloride | Carcinogen; depletion of ozone | Cyclohexane |
| Chloroform | Toxicity; stability | Dimethoxyethane |
| Dichloromethane | Volatile; possible carcinogen | Benzotrifluoride |
| Diethyl ether | Low flash point | Methyl $t$-butyl ether |
| Hexane, pentane | Volatility | Heptane |
| THF | Miscibility with $H_2O$ | 2-MeTHF |
| Dioxane, dimethoxyethane | Toxicity | 2-MeTHF |
| HMPA | Carcinogen | DMPU |

DMPU, $N,N$-Dimethylpropylene urea or 1,3-Dimethyl-3,4,5,6-tetrahydro-2(1$H$)-pyrimidinone; HMPA, hexamethylphosphoric triamide; THF, tetrahydrofuran.

## 1.2.2 Cooling Baths

Often it is necessary to use cooling baths to chill a reaction below 0 °C (Table 1.4). The low temperature could be important for a variety of reasons, including stereoselectivity dependence, decomposition of reagents, and decomposition of products. Thus, a simple and inexpensive method for maintaining the reaction is with a cooling bath. A cooling bath can be produced by creating a slush with solvent and ice, solid $CO_2$ (dry ice), or liquid nitrogen in a Dewar flask. To maintain the temperature, additional amounts of ice, dry ice, or liquid nitrogen must added. *However, an internal thermometer should be used to determine the actual temperature of the reaction mixture.*

Reference: Phillips, A. M.; Hume, D. N. *J. Chem. Edu.* **1968**, *54*, 664; Armarego, W. L. F.; Perrin, D. D. *Purification of Laboratory Chemicals*; 4th ed.; Butterworth-Heineman: Oxford, 1996; p. 36.

Table 1.4 Cooling Baths

| Mixture | Temperature (°C) |
|---|---|
| Crushed ice | 0 |
| Ice/sodium chloride | −5 to −20 |
| Ethylene glycol/$CO_2$(s) | −11 |
| Carbon tetrachloride/$CO_2$(s) | −23 |
| Heptan-3-one/$CO_2$(s) | −38 |
| Acetonitrile/$CO_2$(s) | −41 |
| Chloroform/$CO_2$(s) | −61 |
| Ethanol/$CO_2$(s) | −72 |
| Acetone/$CO_2$(s) | −78 |
| Ethyl acetate/$N_2$(l) | −84 |
| Methanol/$N_2$(l) | −98 |
| Ethanol/$N_2$(l) | −116 |
| Pentane/$N_2$(l) | −131 |

10    MODERN ORGANIC SYNTHESIS IN THE LABORATORY

Table 1.5  Common commercially available Grignard reagents

| Organomagnesium reagent | Solvent (concentration in M) |
|---|---|
| Methylmagnesium chloride | THF (3.0) |
| Methylmagnesium bromide | butyl ether (1.0), ether (3.0), PhMe/THF (75/25, 1.4) |
| Methylmagnesium iodide | ether (1.0) |
| Ethylmagnesium bromide | TBME (1.0), ether (3.0), THF (1.0) |
| Propylmagnesium chloride | Ether (2.0) |
| Isopropylmagnesium | Butyl diglyme (1.4), ether (2.0), THF (2.0) |
| *tert*-Butylmagnesium chloride | Ether (2.0), THF (1.0) |
| Vinylmagnesium chloride | THF (1.0, 1.6) |
| Allylmagnesium bromide | Ether (1.0 ) |
| Allylmagnesium chloride | THF (2.0 ) |
| Phenylmagnesium bromide | Ether (3.0), THF (1.0) |
| Phenylmagnesium chloride | THF (2.0) |

TBME, *tert*-butylmethyl ether; THF, tetrahydrofuran.

*Source*: Sigma-Aldrich. See Sigma-Aldrich ChemFiles 2002, Vol. 2, No.5, for more organomagnesium reagents.

### 1.2.3  Preparation and Titration of Grignard Reagents

A.  Preparation of Grignard Reagents

*Tricks to initiate the Grignard reagent formation*

Typically, dibromoethane or iodine is added to the reaction flask containing the magnesium metal in the solvent. Dibromoethane ($BrCH_2CH_2Br$) reacts with the oxidized magnesium surface, giving $MgBr_2$ and ethylene gas. As a result, fresh magnesium metal is exposed to the organohalide which leads to Grignard reagent formation. Likewise, $I_2$ reacts with the oxidized surface layer, giving $MgI_2$ and exposing fresh magnesium metal to the organohalide. Also, Grignard reagents must be generated in ethereal solvents such as diethyl ether or THF because the reagents form stable chelates with these solvents.

1. Preparation by reaction with magnesium metal

$$Br\text{-}CH(CH_3)\text{-}CH_2\text{-}OBn \xrightarrow[\text{reflux, 30 min.}]{\text{Mg, THF}} BrMg\text{-}CH(CH_3)\text{-}CH_2\text{-}OBn$$

Magnesium turnings (3.5 g, 146 mmol, flamed dried, and cooled under argon) were immersed in THF (10 mL). Two drops of dibromoethane was added to remove the magnesium oxide on the surface of the magnesium turnings. (2*R*)-(−)-Benzyloxy-2-methylpropylbromide (10.4 g, 43 mmol) in THF (70 mL) was added at a rate to keep a gentle reflux (not more than approximately 25% of the bromide should be added before the reaction begins). The reaction mixture was then stirred at reflux for 30 min, cooled to room temperature, and additional THF (40 mL) was added to give a solution of 0.5 M Grignard reagent solution.

Reference: Li, J. *Total Synthesis of Myxovirescin A and Approaches Toward the Synthesis of the A/B Ring System of Zoanthamine*. Ph.D. Thesis, Indiana Univiversity: Bloomington, Indiana; 1996.

## 2. Preparation by reaction with commercially available Grignard reagents

$$\text{EtO}_2\text{C}\underset{S}{\overset{Br}{\diagup\!\!\!\diagdown}}\text{Br} \quad\xrightarrow[\text{2. PhCHO}]{\text{1. }i\text{-PrMgBr}}\quad \text{EtO}_2\text{C}\underset{S}{\overset{Br}{\diagup\!\!\!\diagdown}}\underset{\text{OH}}{\overset{}{\text{CH}(\text{Ph})}}$$
(83% yield)

A solution of *i*-PrMgBr (1.05 mmol) in THF (0.8 M, 1.31 mL) was added dropwise over 5 min to a stirred solution of the starting material (314 mg, 1 mmol) in THF (10 mL) at −40 °C under argon. The resulting solution was then stirred for 30 min, and benzaldehyde (122 μL, 1.20 mmol) was added. The reaction mixture was allowed to warm to room temperature, brine (20 mL) was added, and the reaction was worked up as normal. The crude residue was purified by column chromatography on silica (pentane/Et$_2$O, 4:1) to give the alcohol product (283 mg, 83%) as a colorless oil.

Reference: Abarbri, M.; Thibonnet, J.; Berillon, L.; Dehmel, F.; Rottlander, M.; Knochel, P. *J. Org. Chem.* **2000**, *65*, 4618–4634.

Also see Chapter 5 for additional examples of organomagnesium generation. For more on the preparation of organomagnesium reagents see Wakefield, B. J. *Organomagnesium Methods in Organic Synthesis*; Academic Press: San Diego, CA, 1996, pp. 21–71.

### B. Titration of Grignard Reagents

An effective method for the titration of Grignard reagents is with menthol in the presence of 1,10-phenanthroline in THF. The endpoint of the titration is reached when a violet or burgundy color persists, which is indicative of a charge transfer complex formed beween the organomagnesium reagent and 1,10-phenanthroline.

A 50-mL, flame-dried, one-necked, round-bottomed flask fitted with a magnetic stirring bar was rapidly charged with menthol (312 mg, 2 mmol) and 1,10-phenanthroline (4 mg, 0.02 mmol) before being capped with a rubber septum and flushed with dry nitrogen via a syringe needle. Dry THF (15 mL, distilled from CaH$_2$) was introduced, and the resulting solution was stirred at room temperature under the nitrogen atmosphere. The Grignard solution was then added dropwise by the syringe technique until a distinct violet or burgundy color persisted for longer than a minute.

$$[\text{RMgX}] \text{ in M} = \text{mmol of menthol/volume of RMgX in mL}$$

*Note*: *the volume of RMgX solution required to generate the charge-transfer complex with 1,10-phenanthroline, after formation of the magnesium alkoxide, is negligible and not included in the titer calculation.*

Reference: Lin, H.-S.; Paquette, L. A. *Synthetic Communications* **1994**, *24*, 2503–2506. See also, Watson, S. C.; Eastham, J. F. *J. Organometal. Chem.* **1967**, *9*, 165–168.

## 1.2.4 Preparation and Titration of Common Organolithium Reagents

Typical solvents used for organolithium reactions include THF, diethyl ether (ether), dimethoxyethane, toluene, and hexane. Although THF and diethyl ether are the most common solvents involving organolithium reagents, one should be aware that both

**Table 1.6** Common commercially available organolithium reagents

| Organolithium reagent | Solvent (concentration) |
|---|---|
| $n$-Butyllithium | Cyclohexane (2.0 M); hexanes (1.6, 2.5, 10 M); pentane (2.0 M) |
| $sec$-Butyllithium | Cyclohexane (1.4 M) |
| $tert$-Butyllithium | Pentane (1.7 M) |
| Methyllithium | Diethoxymethane (3.0 M); diethylether (1.6 M) |
| Ethyllithium | Benzene/cyclohexane (90/10, 0.5 M) |
| Phenyllithium | di-$n$-Butylether (2.0 M) |

*Source*: Sigma-Aldrich.

solvents are susceptible to decomposition. THF can undergo metallation at the α-position; the anion can then breakdown through a fragmentation process to deliver the enolate of ethanal and ethylene gas (Wakefield, B. J. *Organolithium Methods*; Academic Press: San Diego, CA, 1988, p. 178). In addition, diethyl ether can decompose to form lithium ethoxide and ethylene gas.

To help understand the stability of some common alkyllithium reagents, Table 1.7 lists the half-lives of $n$-BuLi, $s$-BuLi, and $t$-BuLi in ether and THF.

A. Preparation of Organolithium Reagent

**Table 1.7** Half-lives ($t_{1/2}$) of common alkyllithium reagents in ether and THF at various temperatures

| Alkyllithium/solvent | –40 °C | –20 °C | 0 °C | 20 °C |
|---|---|---|---|---|
| $n$-BuLi/ether | — | — | — | 153 h |
| $n$-BuLi/THF | — | — | 17.3 h | 1.78 h |
| $s$-BuLi/ether | — | 19.8 h | 2.32 h | — |
| $s$-BuLi/THF | Stabilization at 0.4 [$s$-BuLi]$_o$ | 1.30 h | — | — |
| $t$-BuLi/ether | — | 8.05 h | 1.02 h | — |
| $t$-BuLi/THF | 5.63 h | 0.70 h | — | — |

THF, tetrahydrofuran.

*Source*: Stanetty, P.; Mihovilovic, M. D. *J. Org. Chem.* **1997**, *62*, 1514–1515.

A halogen/metal exchange reaction is exemplified here for the preparation of organilithium reagent:

A solution of bis(2-bromophenyl)methane (1.63 g, 5 mmol) in THF (300 mL) was cooled to −30 °C and n-BuLi (2.5 M, 4 mL, 10 mmol) was added dropwise. After completion, the yellow solution was allowed to warm to room temperature and stirred for 2 h. The resulting dilithio reagent was then ready for further reaction.

Reference: Lee, W. Y.; Park, C. H.; Kim, Y. D. *J. Org. Chem.* **1992**, *57*, 4074.

Also see Chapter 5 for additional examples of organolithium generation via halogen/metal exchange. For more on the preparation of organolithium reagents see Wakefield, B. J. *Organolithium Methods*; Academic Press: San Diego, CA, 1988, pp. 21–5.

### B. Titration of Organolithium Reagent

A reliable method for titrating commercial alkyllithium reagents such *n*-BuLi, *sec*-BuLi, and *t*-BuLi is with pivolyl-*o*-toluidine in THF. The first equivalent of the alkyllithium reacts with pivolyl-*o*-toluidine to generate a colorless monoanion. The endpoint of the titration is achieved with the next drop of the alkyllithium solution which metallates the benzylic position of the aryl species to form a yellow to yellow/orange dianion.

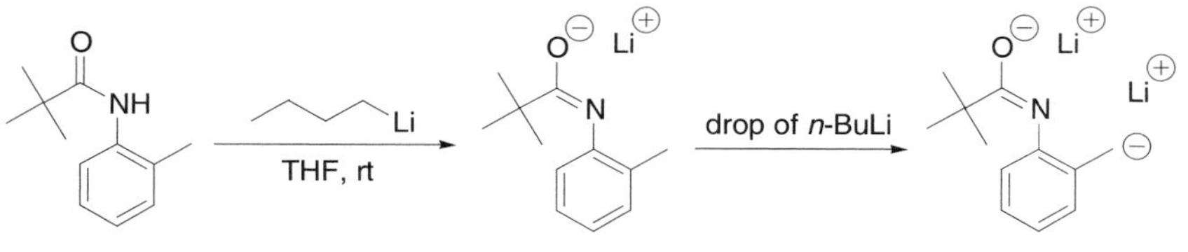

A 25-mL round-bottomed flask fitted with a septum and containing a magnetic stirring bar was evacuated and flushed with argon or nitrogen. Approximately 250–380 mg (0.9–2.0 mmol) of *N*-pivaloyl-*o*-toluidine charged into the flask. Anhydrous THF (5–10 mL) was added, and a white sheet of paper was placed behind the flask. The organolithium solution was added from a 1-mL Hamilton gas-tight syringe. The solution was rapidly stirred under argon. When the titration was completed, the dianion gave an intense yellow color.

[RLi] in M = mmol of *N*-pivaloyl-*o*-toluidine/volume of RLi in mL

*Note*: *the volume of n-BuLi solution required to generate the dianion, after formation of the monoanion, is negligible and not included in the titer calculation.*

Reference: Suffert, J. *J. Org. Chem.* **1989**, *54*, 509.

### 1.2.5 Generation of Zinc Reagent

Organozinc reagents have emerged as a versatile group of organometallic reagents; consequently, the number of reagents available has increased dramatically in recent years. Table 1.8 is a representative subset, but a large number of alkyl- aryl-, and heteroarylorgano zinc reagents are available. For a review of functionalized zinc reagents see Knochel, P.; Millot, N.; Rodriguez, A. L.; Tucker, C. E. *Org. React.* **2001**, *58*, 417–731.

$$\text{I}\smile\smile\overset{\text{CO}_2\text{Bn}}{\underset{\text{NHBoc}}{|}} \xrightarrow[\text{DMF, 0 °C}]{\text{Zn*}} \text{IZn}\smile\smile\overset{\text{CO}_2\text{Bn}}{\underset{\text{NHBoc}}{|}}$$

Zinc dust (325 mesh, 0.147 g, 2.25 mmol, 3 equiv.) was weighed into a 50-mL round-bottomed flask with side arm, which was repeatedly evacuated, heated using a hot air gun, and flushed with nitrogen. Dry DMF (0.5 mL) and trimethylsilyl chloride (6 µL, 0.046 mmol) were added, and the resultant mixture was stirred for 30 min at room temperature. Iodide (0.75 mmol) was dissolved in dry DMF (0.5 mL) under nitrogen. The iodide solution was transferred via a syringe to the zinc suspension at 0 °C, and the mixture was then stirred. Thin layer chromatography (TLC) analysis (petroleum ether–ethyl acetate, 2:1) showed complete consumption of the iodide within 5–60 min.

Reference: Deboves, H. J. C.; Hunter, C. F. W; Jackson, R. F. W. *J. Chem. Soc. Perkin Trans. 1*, **2002**, 733–736.

### 1.2.6 Preparation of Diazomethane

Diazomethane, especially when pure and in large quantities, has been known to be explosive. *Thus, extraordinary caution must be taken to prevent diazomethane from coming into contact with sharp objects. Care must be taken not to allow contact between diazomethane and any metal, ground glass joints, or scratched glassware. In addition, diazomethane solution should not be exposed to direct sunlight or allowed to sit under artificial light for an extended period of time.* For safety concerns see Moore, J. A.; Reed, D. E. *Org. Synth.* **1973**, *Coll. Vol. 5*, 351–354. Diazomethane is never isolated neat, and is typically used in excess (it is a bright yellow solution), and the excess can be quenched with acetic acid to form the easily removed methyl acetate. A safer alternative to diazomethane is the commercially available trimethylsilyldiazomethane (TMSCHN$_2$). TMSCHN$_2$ is sold as a solution in hexanes. See Chapter 2 for this reagent.

EXAMPLE 1

Table 1.8 Some common commercially available zinc reagents

| Organozinc reagents (0.5 M in THF) |
|---|
| Propylzinc bromide |
| Butylzinc bromide |
| Cyclohexylzinc bromide |
| 3-Ethoxy-3-oxopropylzinc bromide |
| Phenylzinc bromide |
| 2-Pyridylzinc bromide |
| 2-Thienylzinc bromide |

THF, tetrahydrofuran.

See Sigma-Aldrich ChemFiles 2002, Vol. 2, No.5.

Diazomethane was generated with a diazomethane-generating glassware kit (Aldrich). A solution of *N*-methyl-*N*-nitroso-4-toluenesulfonamide (Diazald, 2.23 g, 10.4 mmol) in ether (24 mL) was added dropwise to a mixture of KOH (1.75 g, 31.2 mmol) in H$_2$O (18 mL), ether (4 mL), and 2-(2-ethoxyethoxy)ethanol (18 mL) kept at 70 °C. The ethereal solution of diazomethane was continuously distilled into a flask that was ready to use for the next reaction.

Reference: Tchilibon, S.; Kim, S.-K.; Gao, Z.-G.; Harris, B. A.; Blaustein, J. B.; Gross, A. S.; Duong, H. T.; Melman, N.; Jacobson, K.A. *Bioorg. Med. Chem.* **2004**, *12*, 2021–2034.

EXAMPLE 2

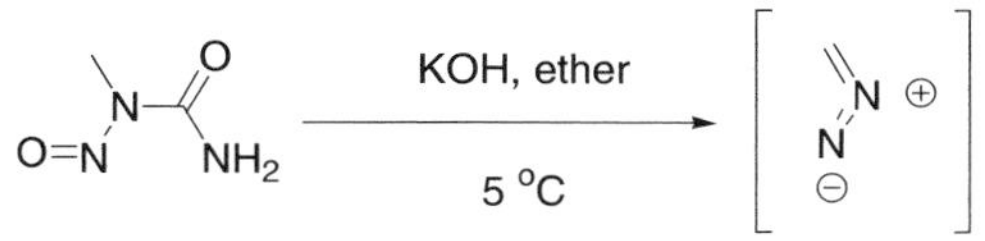

An aqueous solution of potassium hydroxide (40%, 30 mL) was added to diethylether (100 mL), and the mixture was cooled to 5 °C. Finely powdered *N*-nitroso-*N*-methylurea (10 g) was added in small portions over a period of 1–2 min. The deep yellow ether layer can be decanted readily; it contains about 2.8 g of diazomethane together with some dissolved impurities and water. *Note*: *DO NOT use a glass container with a ground glass joint; however, an Erlenmeyer flask or a test tube is usually suitable. In addition, the vessel should be free of any interior scratches. The ethereal layer may be removed using a flame-polished pipette.*

Reference: Arndt, F. *Org. Synth.* **1943**, *Coll. Vol. II*, 165–167.

### 1.2.7 Preparation of the Dess–Martin Reagent

The Dess–Martin oxidation is the method of choice for the oxidation of alcohols bearing sensitive functional groups.

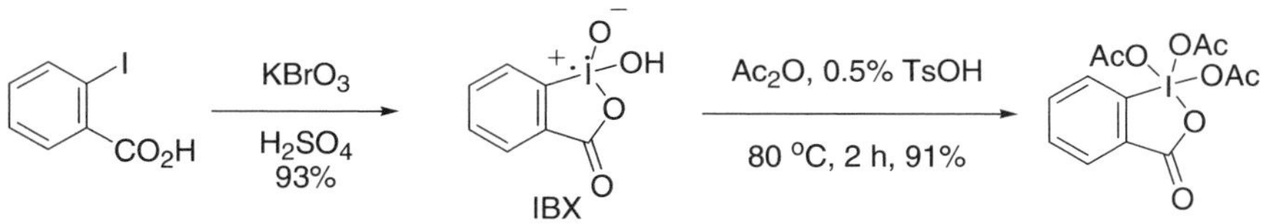

CAUTION! The Dess–Martin precursor [1-hydroxy-l,2-benziodoxol-3(1*H*)-one (IBX)] was reported to be explosive under excessive heating (> 200 °C) or impact. Sporadically, IBX did not decompose explosively at 233 °C, but melted with browning. However, this *cannot be taken as an indication of absence of explosivity as the same batch showed inconsistent results*. An analytically pure sample (≥ 99%) was subjected to explosibility tests.

A. 1-Hydroxy-l,2-benziodoxol-3(1*H*)-one (IBX). (≥ 95% Purity)

2-Iodobenzoic acid (50.0 g, 0.20 mol) was added all at once to a solution of Oxone (181.0 g, 0.29 mol, 1.3 equiv.) in deionized water (650 mL, 0.45 M) in a 2 L flask. The reaction mixture was warmed to 70–73 °C over 20 min and mechanically stirred at this temperature for 3 h. The aspect of the mixture varies consistently during the reaction. The initial thick slurry coating the walls of the flask eventually becomes a finely dispersed, easy to stir suspension of a small amount of solid that sediments easily on stopping the stirring. The suspension was then cooled to 5 °C and left at this temperature for 1.5 h with slow stirring. The mixture was filtered through a medium porosity sintered-glass funnel, and the solid was repeatedly rinsed with water (6 × 100 mL) and acetone (2 × 100 mL). The white crystalline solid was left to dry at room temperature for 16 h and weighed 44.8–45.7 g (79–81%).

Mother and washing liquors were oxidizing and acidic. They were treated with solid $Na_2SO_3$ (70 g, 0.55 mol) and neutralized with NaOH (1 M) before disposal. The internal temperature rose to 30 °C.

Reference: Frigerio, M.; Santagostino, M.; Sputore, S. *J. Org. Chem.* **1999**, *64*, 4537–4538.

B. 1,1,1-Triacetoxy-1,1-dihydro-1,2-benziodoxol-3(1*H*)-one (the Dess–Martin reagent)

IBX 100 g was added to a 1-L round-bottomed flask containing $Ac_2O$ (400 mL), $TsOH \cdot H_2O$ (0.5 g), and a magnetic stirring bar. The flask was equipped with a drying tube and was immersed in an oil bath at ca. 80 °C. The mixture was stirred for 2 h and then cooled in an ice-water bath. The cold mixture was filtered through a fritted glass funnel followed by rinsing with anhydrous ether (5 × 50 mL). The resulting white crystalline solid (138 g, 91% for two steps) was quickly transferred to an argon flushed amber-glass bottle and stored in a freezer: mp 134 °C.

Reference: Ireland, R. E.; Liu, L. *J. Org. Chem.* **1993**, *58*, 2899. See also, Meyer, S. D.; Schreiber, S. L. *J. Org. Chem.* **1994**, *59*, 7549–7552.

### 1.2.8 Preparation of Lithium Diisopropylamide

Lithium diisopropylamide (LDA) is a strong, non-nucleophilic base that is widely used. Because of its reactivity with many solvents, it has historically been generated in the laboratory by reaction of *n*-butyllithium and diisopropyl amine. Recently, LDA has become commercially available as a solution in THF/heptane/ethyl benzene. The commercially available material is frequently colored, making titration difficult.

Freshly prepared LDA has varying stability, being most stable in alkanes and 1:1 alkanes:THF. Homemade LDA should be stored cold to extend its shelf-life. The preparation of LDA is representative of other lithium amide bases, such as lithium tetramethylpiperidide and lithium hexamethyl disilylamide.

To a solution of diisopropylamine (3.44 g, 4.76 ml, 0.0341 mole) in THF (25 mL) at −78 °C (methanol–dry ice bath) is added a solution of *n*-butyllithium (1.61 M in hexane, 21.1 mL, 0.0340 mol) with stirring under argon. The solution is warmed to 0 °C in 15 min to provide an approximately 0.7 M solution of LDA.

Reference: Enders, D.; Pieter, R.; Renger, B.; Seebach, D. *Org. Synth.* **1978**, *58*, 113 or Enders, D.; Pieter, R.; Renger, B.; Seebach, D. *Org. Synth.* **1988**, *Coll. Vol. 6*, 542.

### 1.2.9 Jones Reagent

The Jones reagent is prepared by first dissolving chromium trioxide (70 g, 0.70 mol) in water (100 mL) in a 500 mL beaker. The beaker is then immersed in an ice bath,

18 M sulfuric acid (61 mL, 1.10 mol), and water (200 mL) is added cautiously with manual stirring. The solution is cooled to 0–5 °C.

Reference: Meinwald, J.; Crandall, J.; Hymans, W. E. *Org. Syn.* **1973**, *Coll. Vol. 5*, 866.

## 1.3 Chromatography

In this section, two types of chromatography are presented: TLC and flash chromatography.

### 1.3.1 Thin Layer Chromatography

Reference: Wall, P. E. *Thin Layer Chromatography*; Royal Society of Chemistry: Cambridge (UK), 2005.

The thin layer chromatography (TLC) plates that we buy are slide glasses coated with silica gel. In order to visualize UV-active organic compounds, the silica gel is coated with a layer of fluorescent dye "Flur." Therefore, UV-active compounds can be detected under UV light. Occasionally, non-UV-active compounds are also seen. One case is iodides, which are UV-light quenchers, while another case is inorganic salts, often seen on the baseline; the salts are visualized under UV simply because they cover the fluorescent dye "Flur."

The TLC plates that we buy are somewhat too large. It is a good idea to cut the plates with a diamond-type or carbide roller into smaller plates. Not only does this save costs, it is also faster to develop a smaller TLC plate than a larger one.

Retention factor, $R_f$, is a measurement of how far up a plate the compound travels. $R_f$ is calculated by *the distance of the center of the spot from the baseline (P or SM)* divided by *the distance of the solvent front from the baseline (S)*. Column volume (CV) is $1/R_f$ (Figure 1.3). These values will be referenced in the discussion of column chromatography.

In general, the eluting strength of commonly used solvents for normal phase chromatography is: stationary phase = silica gel neutral alumina; increasing order proceeds as follows: petroleum ethers < hexanes < cyclohexane < toluene < diethyl ether < dichloromethane < chloroform < ethyl acetate < acetone < ethanol < methanol < acetic acid.

Another useful technique is two-dimensional TLC (Figure 1.4). This method is a simple way of determining if decomposition of a desired product or products will occur on the stationary phase (typically silica gel or alumina). After spotting your compound on the TLC plate and developing the plate, the plate is turned 90° and then developed again. For a single component, one spot should be visible with the same $R_f$ as the spot from the first development. If streaking is observed in the second development and/or new spots arise, decomposition has occurred. For compound mixtures, the final developed spots will fall in a straight line if no decomposition has occurred; however, if the spots do not fall in a straight line or additional spots appear within a lane, decomposition occurred during TLC development. Decomposition is usually an indication that a chemical interaction occurred with the stationary phase.

### 1.3.2 Recipes of Common Thin Layer Chromatography Stains

For compounds that do not visualize well under UV, TLC stains are necessary to visualize the spots. Listed herein are some popular TLC stains often used in an

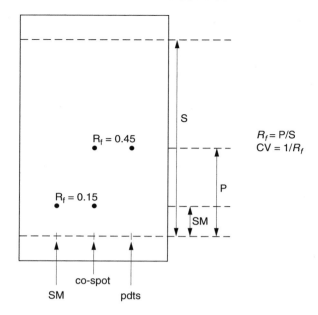

Figure 1.3 One-dimensional TLC.

organic chemistry laboratory. Most of these visualization stains require the submersion of the TLC plate in a solution of stain, carefully blotting excess stain, and then heating the stained TLC plate on a hot plate or with a heat gun. *All of these manipulations should be performed in a well ventilated hood, particularly the heating step, because many of these stains are toxic!*

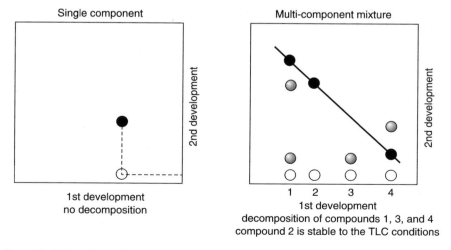

Figure 1.4 Two-dimensional TLC.

### p-Anisaldehyde Stain

The greatest advantage of this stain is that different colors are manifested on TLC on heating for different molecules. Therefore, molecules can be differentiated even if they have the same $R_f$ values. The disadvantage is its strong but pleasant odor release during heating (*toxic, in the hood!*).

| | |
|---|---|
| p-Anisaldehyde | 12 g (11 mL) |
| Ethanol | 500 mL |
| Concentrated $H_2SO_4$ | 5 mL |

### Cerium Sulfate Stain

General stain. Most compounds are stained brown or yellow.

Cerium Sulfate (8 g)

15% sulfuric acid (100 mL)

### Hanessian Stain (cerium molybdate stain)

One of the most sensitive stains which detects most functional groups. The disadvantage is that everything stains blue.

| | |
|---|---|
| $Ce(SO_4)_2$ | 5g |
| $(NH_4)_6Mo_7O_{24} \cdot 4H_2O$ | 25 g |
| Concentrated $H_2SO_4$ | 50 mL |
| $H_2O$ | 450 mL |

### $I_2$ or $I_2$ in Silica Gel

Everything stains yellow. Solid $I_2$ can be added to a developing chamber and the TLC plate developed by placing the plate in the chamber. Alternatively, the $I_2$ can be dispersed on silica gel in a developing chamber, and the TLC plate immersed in the silica gel; the silica gel development tends to be faster. The spots will fade, but the plate can be redeveloped by placing the plate back into the $I_2$ chamber.

### $KMnO_4$

Detects molecules with an "oxidizable" functional group. Relatively insensitive, everything stains yellow, frequently even without heating. Eventually, the entire plate will turn yellow and the spots will be indistinguishable from the background.

| | |
|---|---|
| $KMnO_4$ | 6 g |
| $K_2CO_3$ | 40 g |
| 5% aqueous NaOH | 10 mL |
| $H_2O$ | 600 mL |

### Ninhydrin Stain

Especially sensitive to amino acids, as well as amines and anilines. Avoid contact with skin, or all of your friends will know that you have had an affair with nihydrin.

0.25% of nihydrin in water

*Phosphomolybdic Acid (PMA)*

Everything stains blue–green. A very sensitive stain, possibly used most often. A 20% PMA in ethanol is commercially available; this should be diluted to 5% with ethanol.

| | |
|---|---|
| Phosphomolybdic acid | 12 g |
| Ethanol | 500 mL |

*Vanillin Stain*

Different colors are manifested on heating for different molecules. Smells great, too (*do not inhale intentionally!*).

| | |
|---|---|
| Vanillin | 12 g |
| Ethanol | 250 mL |
| Concentrated $H_2SO_4$ | 50 mL |

### 1.3.3 Flash Chromatography

Reference: Still, W. C.; Kahn, M.; Mitra, A. *J. Org. Chem.* **1978**, *43*, 2923.

One of the most versatile methods of purifying reaction mixtures is flash chromatography using silica gel. Surprisingly, silica gel is one of the most hazardous chemicals in an organic chemistry laboratory. Inhalation of the fine powder of silica gel could cause lung diseases, even lung cancer. Therefore, flash chromatography using silica gel should be carried out in a fume hood. Glass chromatography columns must be inspected routinely for cracks to ensure that the glass is strong enough to withstand the pressure that will be applied. Care must be taken not to apply too much pressure to the column.

Before setting up a column for the separation of the reaction mixture, the type of silica gel to use must be decided. The two most common grades of silica gel are gravity silica gel and silica gel for a flash column. Gravity silica gel, more suitable for easy separations, is composed of larger sized silica gel, 0.062–0.200 mm (70–230 mesh ASTM). In contrast, silica gel for the flash column, more suitable for more difficult separations, is finer, 0.040–0.063 mm (230–400 mesh ASTM). Regarding flash chromatography, Still's flash chromatography technique is the standard procedure. A typical flash chromatography setup is shown in Figure 1.5.

According to Still's 1978 *Journal of Organic Chemistry* paper (Still, W. C.; Kahn, M.; Mitra, A. *J. Org. Chem.* **1978**, *43*, 2923), the size of the glass column depends on the scale of the separation. First, a solvent system is chosen based on its polarity so that the desired product has an $R_f$ of 0.3. For a mixture of 1 g of compounds, 25 g of silica gel is used, on average.

The greatest advantage of Still's flash chromatography technique is its expediency. On average, a column should not take more than 10–15 min. The rule of thumb is, if it takes more than 30 min for a routine chromatographic separation, it is likely to be unsuccessful or the flow rate is too slow. The trick is to select a column of appropriate diameter. The height of the silica gel should not be taller than 6 inches, changing the diameter to accommodate the endpoint, where 25 g of silica gel is loaded for each 1 g of the reaction mixture to be separated. For more difficult separations, a ratio of 100:1 to 200:1 silica gel to mass of mixture is not unusual. As a consequence, run times are extended.

In the example in Figure 1.3, the product will come out after 2.2 (1/0.45) column volumes, while the starting material will come out after 6.7 column volumes (1/0.15). The column volume is approximately 1.25 times the mass of silica gel used (e.g., a column with 2 g of silica gel has a column volume of approximately 2.5 mL). In this case, a more polar solvent system could be developed to make the compounds come off the silica gel more quickly, or a silica gel:compound ratio of less than the 25:1 could be used. The fraction size should be approximately twice the column volume; if $\Delta$CV is < 1, this will obviously have to be smaller.

On the other hand, although it is a good idea to start the solvent system to ensure the $R_f$ values range from 0.2 to 0.3, it is even better to apply a gradient solvent system by increasing the polarity. For instance, gradually changing the solvent system from 10% EtOAc/hexane to 50% EtOAc/hexane will afford a much faster separation than running the column at a constant 10% EtOAc/hexane solvent system. For very polar compounds, small quantities of aqueous ammonium hydroxide or triethyl amine (usually < 5% by volume) are added to a binary eluent such as $CH_2Cl_2$/MeOH to decrease "streaking" of amines down the column. The same argument holds true when purifying acids, although small quantities of acetic acid are used in the binary solvent mixture.

In addition to traditional flash column chromatography, there are now commercially available low and medium pressure chromatography systems. Manufacturers

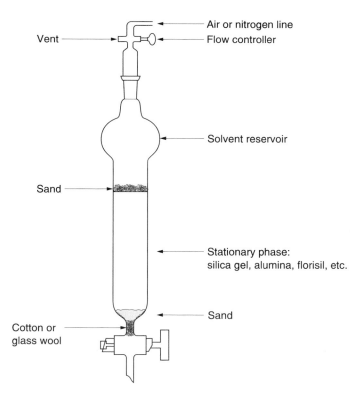

Figure 1.5 Typical flash chromatography setup.

include Analogix, Biotage, Teledyne ISCO, and many others (see Figure 1.6). The sophistication of these commercially available instruments varies and this is reflected in cost. An effective yet relatively inexpensive setup involves a solvent reservoir made of stainless steel and a commercial silica gel column housed in a metal chamber (i.e., Analogix F12/40 System™). The solvent is pushed from the reservoir via air or nitrogen through the column. The eluent is then collected in test tubes, as seen in Figure 1.6b. More elaborate systems are computer-driven which include interactive software, an electric pump/solvent mixer, UV detector, and automated fraction collector. These systems allow for tremendous flexibility and advantages over the traditional flash chromatography (i.e., Teledyne-Isco CombiFlash®Companion®). They include prepacked columns, high flow rates with good to excellent resolution of solutes, shorter run times, "true" solvent gradients, fraction monitoring by UV, and manipulation of the chromatographic conditions during the purification run. In addition, prepacked reverse phase columns (C18) are available. These instruments can separate samples in the milligram to kilogram range. Larger scale purification systems are also available that purify kilograms of material (i.e., Biotage Flash 400™).

## 1.4 Crystallization

Crystallization is the most desirable method of purification on a large scale for solids. Crystallizations require less solvent, time, and resource once they have been developed. Even above a 10 g scale, a little time invested on crystallization can save hours in front of a rotary evaporator.

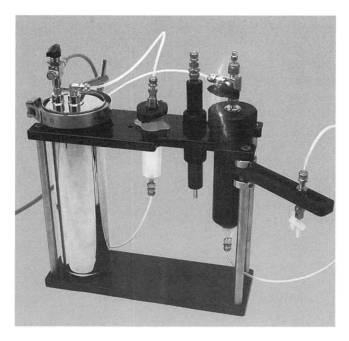

Figure 1.6 Commercial flash chromatography setups. (A) Flash 12/40 System™ (reprinted with permission from Analogix).

The ideal situation is a single solvent crystallization (Table 1.9). The compound of interest is not soluble in a solvent at low temperatures but very soluble at higher temperatures. In cases such as these, the material should be dissolved in a minimum amount of hot solvent and the solution allowed to cool slowly with stirring.

A good first pass approach in cases where a compound will not crystallize from a single solvent is to find a solvent in which the compound of interest is relatively soluble and a solvent miscible in the first in which the compound is not soluble (Table 1.10). Dissolve the compound in a minimum amount of the first solvent near the lower of the two boiling points, and then slowly add a second solvent, in which the compound is not soluble, until the solution becomes cloudy. If the antisolvent has a much lower boiling point than the solvent, allow the solution to cool slowly with stirring and add seeds periodically until the solids persist.

Very impure compounds typically do not crystallize well. Additionally, the target compound can oil out of solution if the solution reaches saturation above the melting point. Crystallizations for purification should be stirred. Crystallizations benefit by seeding with crystalline material; therefore, it may be beneficial to purify a small amount of material by column chromatography to determine the melting point, and to have seeds before proceeding with a crystallization.

Growing crystals for structure determination relies on similar principles, but requires slower crystal growth. If the crystals grown as described above are not sufficiently large for X-ray analysis, they have grown too fast. To slow the growth, slow evaporation can be attempted, where the compound is dissolved in an excess of

Figure 1.6 (*cont.*) (B) CombiFlash®Companion® (reprinted with permission from Teledyne-ISCO).

Figure 1.6 (*cont.*) (C) Flash 400™ (reprinted with permission from Biotage).

solvent (either single solvent or mixture) and that solvent is allowed to slowly evaporate. Alternatively, a solution of the compound in a vial can be placed in a larger flask containing an antisolvent and capped (Figure 1.7). The solvents will slowly exchange, resulting in crystallization.

## 1.5 Residual Solvent Peaks in Nuclear Magnetic Resonance

One of the daily nuisances of an organic chemist is interpreting residual solvent peaks in NMR spectra. One trick is to dissolve the compound with the solvent whose deuterated solvent one is going to use for taking the NMR spectrum. For instance, if $CDCl_3$ is the solvent being used for taking the NMR spectrum, $CHCl_3$ can be employed to

# FUNDAMENTAL TECHNIQUES 25

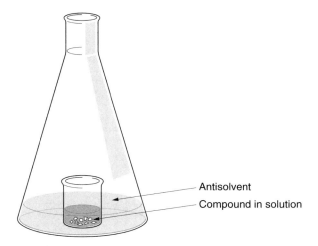

Figure 1.7 Crystallization setup.

dissolve the product and remove it in vacuo. After repeating the process a few times, most volatile residual solvents will be gone and a clean spectrum can be obtained.

As $CDCl_3$ is one of the most popular deuterated solvents used in the organic laboratory, the chemical shifts in $^1H$ NMR spectra of the most common laboratory solvents as trace impurities are in Table 1.11.

It is often the case that the chemical shifts of the same compounds are not the same in different deuterated solvents. To that end, $d_6$-DMSO $^1H$ NMR data are listed in Table 1.12.

Table 1.9 Common crystallization solvents

| Solvent | Boiling point (°C)[†] | Dielectric constant[†] |
|---|---|---|
| Diethyl ether | 34.5 | 4.27 |
| Dichloromethane | 40 | 8.93 |
| Acetone | 56 | 21.01 |
| Chloroform | 61 | 4.81 |
| Methanol | 64.6 | 33.0 |
| Tetrahydrofuran | 65 | 7.52 |
| Hexane | 68.7 | 1.89 |
| Ethyl acetate | 77.1 | 6.08 |
| Ethanol | 78.2 | 25.3 |
| Benzene* | 80.0 | 2.28 |
| Acetonitrile | 81.6 | 36.64 |
| Isopropyl alcohol | 82.3 | 20.18 |
| Heptane | 98.5 | 1.92 |
| Water | 100 | 80.10 |
| 1,4-Dioxane | 101.5 | 2.22 |
| Toluene | 110.6 | 2.38 |
| Acetic acid | 117.9 | 6.20 |
| Butyl acetate | 126.1 | 5.07 |
| N,N-dimethylformamide | 153 | 38.25 |

[†]Lide, D. R., Ed. *Handbook of Chemistry and Physics*, 84th ed.; CRC Press LLC: Boca Raton, FL, 2003, pp. 8–129.
*Toluene is an excellent substitute for benzene.

## Table 1.10 Miscible solvents

| | |
|---|---|
| Diethyl ether | Acetone, ethyl acetate, ethanol, methanol, acetonitrile, hexanes |
| Dichloromethane | Acetic acid, acetone, toluene, ethyl acetate, methanol, isopropyl alcohol |
| Acetone | Toluene, dichloromethane, ethanol, ethyl acetate, acetonitrile, chloroform, hexanes, water |
| Chloroform | Acetic acid, acetone, toluene, ethanol, ethyl acetate, hexanes, methanol |
| Methanol | Dichloromethane, chloroform, ethyl ether, water |
| Hexanes | Toluene, dichloromethane, chloroform, ethanol |
| Ethyl acetate | Acetic acid, methanol, acetone, chloroform, dichloromethane |
| Ethanol | Acetic acid, acetone, chloroform, dichloromethane, diethyl ether, hexanes, toluene, water |
| Toluene | Acetone, diethyl ether, chloroform, dichloromethane, ethanol, acetonitrile |
| Water | Acetic acid, acetone, ethanol, methanol, acetonitrile |

Adapted from Armarego, W. L. F.; Perrin, D. D. *Purification of Laboratory Chemicals,* 4th ed.; Butterworth-Heinemann: Oxford, **1996**, p.35.


## Table 1.11 $^1$H NMR data of the most common laboratory solvents as trace impurities in CDCl$_3$ (in ppm)

| Solvent | Pattern, chemical shift ($\delta$) | Pattern, chemical shift ($\delta$) | Pattern, chemical shift ($\delta$) |
|---|---|---|---|
| Chloroform | s, 7.26 | — | — |
| Acetic acid | s, 2.10 | — | — |
| Acetone | s, 2.10 | — | — |
| *t*-Butyl methyl ether | s, 1.19 | s, 3.22 | — |
| Dichloromethane | s, 5.30 | — | — |
| Diethyl ether | t, 1.21 | q, 3.48 | — |
| Dimethylformaldehyde | s, 8.02 | s, 2.96 | s, 2.88 |
| Dimethylsulfoxide | s, 2.62 | — | — |
| et alhyl acetate | q, 4.12 | s, 2.05 | t, 1.26 |
| Grease | br s, 1.26 | m, 0.86 | — |
| *n*-Hexane | m, 1.26 | t, 0.88 | — |
| Methanol | s, 3.49 | s, 1.09 | — |
| Pyridine | m, 8.63 | m, 7.68 | m, 7.29 |
| Silicon grease | s, 0.07 | — | — |
| Tetrahydrofuran | m, 3.76 | m, 1.85 | — |
| Toluene | s, 7.19 | s, 2.34 | — |
| Triethylamine | t, 2.53 | t, 1.03 | — |
| Water | s, 1.60 | — | — |

Reference: Gottlieb, H. E.; Kotlyar, V.; Nudelman, A. *J. Org. Chem.* **1997**, *62*, 7512–7515.

Table 1.12 $^1$H NMR data of the most common laboratory solvents as trace impurities in $d_6$-DMSO (in ppm)

| Solvent | Pattern, chemical shift (δ) | Pattern, chemical shift (δ) | Pattern, chemical shift (δ) |
|---|---|---|---|
| Acetic acid | s, 1.91 | — | — |
| Acetone | s, 2.09 | — | — |
| *t*-Butyl methyl ether | s, 1.11 | s, 3.08 | — |
| Chloroform | s, 8.32 | — | — |
| Dichloromethane | s, 5.76 | — | — |
| Diethyl ether | t, 1.09 | q, 3.38 | — |
| Dimethylformaldehyde | s, 7.95 | s, 2.89 | s, 2.73 |
| Dimethylsulfoxide | s, 2.54 | — | — |
| Ethyl acetate | q, 4.03 | s, 1.99 | t, 1.17 |
| Grease | — | — | — |
| *n*-Hexane | m, 1.25 | t, 0.86 | — |
| Methanol | s, 4.01 | s, 3.16 | — |
| Pyridine | m, 8.58 | m, 7.79 | m, 7.39 |
| Silicon grease | — | — | — |
| Tetrahydrofuran | m, 3.60 | m, 1.76 | — |
| Toluene | s, 7.18 | s, 2.30 | — |
| Triethylamine | t, 2.43 | t, 0.93 | — |
| Water | s, 3.33 | — | — |

Reference: Gottlieb, H. E.; Kotlyar, V.; Nudelman, A. *J. Org. Chem.* **1997**, *62*, 7512–7515.

# 2

# Functional Group Manipulations

## 2.1 Alcohol Oxidation State

### 2.1.1 Alcohols

#### 2.1.1.1 Alkyl Alcohol to Alkyl Bromide

$CBr_4$–$Ph_3P$ is very straightforward and widely used. Workup and purification can be messy at times because of the by-product, $Ph_3PO$.

1.2 eq. $CBr_4$, 1.5 eq. $Ph_3P$
$CH_2Cl_2$, 0 °C to rt, 1 h
93%

To a mixture of the alcohol (0.800 g, 3.36 mmol) and carbon tetrabromide (1.337 g, 4.03 mmol) in $CH_2Cl_2$ at 0 °C was added a solution of $PPh_3$ (1.319 g, 5.03 mmol) in $CH_2Cl_2$ (3 mL). The reaction mixture was stirred at room temperature for 1 h, concentrated under reduced pressure, and purified by column chromatography to afford the bromide (0.941 g, 93% yield).

Reference: Hu, T.-S.; Yu, Q.; Wu, Y.-L.; Wu, Y. *J. Org. Chem.* **2001**, *66*, 853–861.

A two-step sequence consisting of mesylate formation followed by treatment with LiBr can also be used. This procedure involves two steps, but workup and purification are very straightforward. The bromide can be carried out to the next step without further purification in many cases.

To a solution of 5-hydroxymethyl-1-methylcyclopentene (3.8 g, 34 mmol) in $CH_2Cl_2$ (50 mL) at 0 °C was added triethylamine (5.2 mL, 37 mmol) followed by methanesulfonyl chloride (2.9 mL, 37 mmol). The mixture was stirred at 0 °C for 5 h and then water was added. The organic layer was separated and the aqueous layer was extracted with ether. The combined organic extracts were dried over $MgSO_4$ and the solvent was removed under reduced pressure to give 6.4 g (98%) of (2-methylcyclopent-2-enyl)methyl methanesulfonate, which was used in the next step without further purification.

A solution containing the mesylate (6.4 g, 34 mmol) in acetone (70 mL) was treated with lithium bromide (8.89 g, 102 mmol). The mixture was heated at reflux for 6 h, cooled to room temperature, diluted with water, extracted with ether, and the combined ethereal extracts were dried over $MgSO_4$. Removal of the solvent under reduced pressure gave 4.6 g (78%) of 5-bromomethyl-1-methylcyclopentene, which was used in the next step without further purification.

Reference: Padwa, A.; Dimitroff, M.; Liu, B. *Org. Lett.* **2000**, *2*, 3233–3235.

### 2.1.1.2 Alkyl Alcohol to Alkyl Chloride

$CCl_4$–$Ph_3P$ is a straightforward and widely used method for converting alcohols to alkyl chlorides. Workup and purification can be tricky at times because of the by-product, $Ph_3PO$. An alternative source of $Ph_3P$ is a commercially available triphenylphosphine resin. The advantage of the resin is that the by-product, triphenylphosphine oxide, is bound to the resin, and purification is simplified.

A solution of BOC-sulfamoylaminoalcohol (5.35 mmol), triphenylphosphine (16.05 mmol), and $CCl_4$ (16.05 mmol) in anhydrous acetonitrile (100 mL) was refluxed for 8 h. After cooling to room temperature, the solution was concentrated in vacuo. The residue was triturated with diethyl ether (3 × 150 mL). Triphenylphosphine oxide, which precipitates in the combined organic layers, was removed by filtration. The filtrate was concentrated and the residue purified on silica gel ($CH_2Cl_2$) to afford $N^1$-BOC,$N^3$-(2-chloroethyl) sulfamide with 85% yield.

Reference: Regaïnia, Z.; Abdaoui, M.; Aouf, N.-E.; Dewynter, G.; Montero, J.-L. *Tetrahedron* **2000**, *56*, 381–387.

An alternate procedure involves mesylate formation and $S_N2$ replacement by LiCl. This reaction should be monitored carefully (especially with active alcohols such as allylic and benzylic alcohols) as frequently the chloride by-product will displace the mesylate in situ.

A solution of ($S_s$)-2-(p-tolylsulfinyl)-2-propen-1-ol (1.1 g, 5.6 mmol) in anhydrous dimethylformamide (DMF) (8 mL) under argon was treated at 0 °C with $Et_3N$ (630 mg, 6.2 mmol) and methanesulfonyl chloride (705 mg, 6.2 mmol). The reaction mixture was allowed to warm to room temperature and monitored by thin layer chromatography (TLC) ($CH_2Cl_2$/MeOH 97:3) until the starting alcohol was not detected. The mixture was next diluted with additional anhydrous DMF (10 mL), and LiCl (952 mg, 22.4 mmol) was added portion-wise. Stirring at room temperature was continued until TLC monitoring indicated total consumption of the intermediate mesylate. The reaction mixture was then evaporated to dryness, and the oily residue was taken up in ether and washed with brine. Drying of the organic phase (anhydrous $Na_2SO_4$) and evaporation afforded crude material that was flash-chromatographed (hexanes/EtOAc 85:15) to afford the allylic chloride (0.782 g, 65% yield).

Reference: Márquez, F.; Llebaria, A.; Delgado, A. *Org. Lett.* **2000**, *2*, 547–549.

### 2.1.1.3 Alkyl Alcohol to Alkyl Iodide

$I_2$–$Ph_3P$–Imidazole is a straightforward and widely used method to form alkyl iodides. Workup and purification can be awkward at times because of the by-product, $Ph_3PO$.

Iodine (536 mg, 2.11 mmol) was added to a solution of imidazole (359 mg, 5.27 mmol) and $PPh_3$ (509 mg, 1.94 mmol) in $CH_2Cl_2$ (10 mL) at 0 °C. The solution was stirred for 5 min, and the alcohol (750 mg, 1.76 mmol), dissolved in $CH_2Cl_2$ (2 mL), was then added slowly. The reaction mixture was stirred for 4 h with the exclusion of light. It was then quenched by the addition of an aqueous $Na_2S_2O_3$ solution (10 mL). The aqueous layer was extracted with methyl *tert*-butyl ether (MTBE 3 × 20 mL). The combined organic layers were washed with saturated aqueous NaCl (10 mL) and dried with $MgSO_4$. The solvents were removed in vacuo. The crude product was purified by flash column chromatography (20 g silica, petroleum ether/MTBE 20:1) to yield the iodide (836 mg, 1.54 mmol, 88%) as a colorless oil.

Reference: Arndt, S.; Emde, U.; Bäurle, S.; Friedriech, T.; Grubert, L.; Koert, U. *Chem. Eur. J.* **2001**, 993–1005.

Alkyl alcohols can be converted to alkyl iodides via a two-step sequence consisting of mesylate formation followed by treatment with NaI. This procedure involves two steps, but workup and purification are very straightforward. The iodide can be carried on to the next step without further purification in many cases.

MsCl, Et$_3$N, CH$_2$Cl$_2$ / −10 °C / 90%

NaI, acetone / reflux, 4 h / 71%

To a solution of freshly distilled triethylamine (1.82 g, 18 mmol) in anhydrous CH$_2$Cl$_2$ (35 mL) under argon at −10 °C, the alcohol (2.0 g, 12 mmol) was added dropwise. After stirring for 10 min, methanesulfonyl chloride (1.60 g, 14 mmol) was added. The reaction mixture was stirred for a further 20 min and then washed with saturated aqueous sodium hydrogen carbonate (20 mL), water (20 mL), and brine (10 mL), dried and concentrated in vacuo. The crude mesylate was chromatographed, eluting with petroleum ether:EtOAc (10:1), to obtain the mesylate (2.68 g, 90%) as an off-white waxy solid. To a solution of sodium iodide (0.525 g, 3.5 mmol) in hot anhydrous acetone (30 mL) under argon, was added dropwise a solution of the mesylate (0.5 g, 2 mmol) in anhydrous acetone (25 mL), and the resulting mixture was heated under reflux for 4 h. The acetone was then removed in vacuo and the residue dissolved in ether. Water was added to dissolve the remaining sodium iodide and the organic layer was separated, washed with water (50 mL) and brine (50 mL), dried, and concentrated in vacuo. The residue was chromatographed, eluting with pentane, yielding the iodide (0.396 g, 71%) as a pale brown oil.

Reference: Crombie, B. S.; Smith, C.; Varnavas, C. Z.; Wallace, T. W. *J. Chem. Soc. Perkin Trans.* **2001**, *1*, 206–215.

### 2.1.1.4 Alkyl Alcohol to Azide

Treatment of an alcohol with diphenylphosphoryl azide is a one-step procedure to make azides. But treatment of a mesylate with NaN$_3$ is also frequently used for converting an alcohol to the corresponding azide as a two-step procedure.

The alcohol (1.74 g, 4.99 mmol) was dissolved in toluene (30 mL) and cooled to 0 °C. Diphenylphosphoryl azide (1.65 g, 5.98 mmol) was added followed by 1,8-diazabicyclo[5.4.0]undec-7-ene (0.91 g, 5.98 mmol). The reaction mixture was stirred at 0 °C for 2 h. The solution was then warmed to 25 °C followed by the addition of EtOAc (100 mL). The organic layer was washed sequentially with H$_2$O (1 × 30 mL), saturated NaHCO$_3$ (1 × 50 mL), and brine (1 × 50 mL). The organic layer was then dried over MgSO$_4$, filtered, and concentrated in vacuo. Chromatography on silica gel (7.5% EtOAc/hexanes) provided the azide (1.58 g, 85%) as a light yellow oil.

Reference: Stachel, S. J.; Lee, C. B.; Spassova, M.; Chappell, M. D.; Bornmann, W. G.; Danishefsky, S. J.; Chou, T.-C.; Guan, Y. *J. Org. Chem.* **2001**, *66*, 4369–4378.

### 2.1.1.5 Alkyl Alcohol (Halohydrin) to Epoxide

The chlorohydrin (17.61 g, 53.8 mmol) and $K_2CO_3$ (13.2 g, 88.5 mmol) in MeOH (88 mL) were stirred at room temperature for 3 h. The reaction mixture was diluted with ether (350 mL), washed with water (3 × 80 mL) and brine (100 mL), dried over $Na_2SO_4$, and concentrated in vacuo. The crude product was purified by flash chromatography ($SiO_2$, 20% EtOAc in hexanes) to yield the epoxide as a colorless oil (14.13 g, 97%).

Reference: Gao, L.-X.; Murai, A. *Heterocycles* **1995**, *42*, 745–774.

### 2.1.1.6 Alcohol to Olefin via Burgess Dehydrating Reagent

Burgess dehydrating reagent is efficient at generating olefins from secondary and tertiary alcohols. It does notwork with primary alcohols.

The Burgess reagent (62.0 mg, 0.260 mmol) was added to a solution of the alcohol (19.6 mg, 0.0336 mmol) in 1.2 mL of benzene. The heterogeneous mixture was stirred under nitrogen at ambient temperature for 12 h and then warmed briefly to 50 °C to complete the elimination. After cooling, the mixture was diluted with 5 mL of ether, and 0.5 mL of $H_2O$ was added. The organic layer was removed and dried by passage through a magnesium sulfate plug. Flash chromatography (25% ether in hexanes) furnished the olefinic products (18.2 mg, 96%) as an inseparable 3:1 mixture of isomers favoring the desired trisubstituted olefin.

Reference: Nakatsuka, M.; Ragan, J. A.; Sammakia, T.; Smith, D. B.; Uehling, D. E.; Schreiber, S. L. *J. Am. Chem. Soc.* **1990**, *112*, 5583–5601.

### 2.1.1.7 Alcohol to Olefin via Martin's Sulfurane

Martin's sulfurane dehydrates secondary and tertiary alcohols to give olefins, but forms ethers with primary alcohols.

A solution of Martin's sulfurane (0.949 g, 1.41 mmol) in $CH_2Cl_2$ (5 mL) was slowly added under an atmosphere of nitrogen to a stirred solution of the alcohol (0.41 g, 0.94 mmol) in $CH_2Cl_2$ (10 mL) at room temperature. The resulting pale yellow solution was stirred for 24 h at which time the solvent was removed in vacuo to yield the crude product as a pale yellow oil. Purification by flash column chromatography [$SiO_2$; EtOAc:petroleum ether (40:60)] followed by recrystallization ($CH_3OH/H_2O$) furnished the olefin product as a colorless solid (0.33 g, 84% yield).

Reference: Begum, L.; Box, J. M.; Drew, M. G. B.; Harwood, L. M.; Humphreys, J. L.; Lowes, D. J.; Morris, G. A.; Redon, P. M.; Walker, F. M.; Whitehead, R. C. *Tetrahedron* **2003**, *59*, 4827–4841.

### 2.1.1.8 Mitsunobu Reaction

The venerable Mitsunobu reaction remains a powerful method for the stereoselective inversion of chiral secondary alcohols. When considering this reaction, the chemist should take into account the solvent, the phosphine, the azodicarboxylate, and the nucleophile. The most common solvents used for the Mitsunobu reaction include tetrahydrofuran (THF), dioxane, $CH_2Cl_2$, $CHCl_3$, $Et_2O$, DMF, toluene, and benzene. Also, the two most common azodicarboxylates used are diethylazodicarboxylate (DEAD) and diisopropyldicarboxylate. In terms of the phosphine, most reactions are run in the presence of triphenylphosphine, although removal of the resulting triphenylphosine oxide can be problematic. Moreover, it has been shown that the carboxylic acid nucleophile can play a critical role in this reaction, and therefore some carboxylic acids include formic acid, acetic acid, benzoic acid, 4-nitrobenzoic acid, and 3,5-dinitrobenzoic acid. Finally, in some instances it is advantageous to pre-mix the phosphine with the azodicarboxylate and then add the alcohol. Although the Mitsunobu is most famous for the inversion of chiral alcohols, the scope of this reaction has been expanded to include formation of C–N, C–S, C–halogen, and C–C bonds.

Reviews: (a) Hughes, D. L. *Org. Prep. Proc. Int.* **1996**, *28*, 127–164. (b) Hughes, D. L. *Org. React.* **1992**, *42*, 335–656. (c) Castro, B. R. *Org. React.* **1983**, *29*, 1–162. (d) Mitsunobu, O. Synthesis **1981**, 1–28.

Diethyl azodicarboxylate (11.6 mL, 74.0 mmol) was added dropwise with stirring to a solution of the alcohol (6.20 g, 14.9 mmol), Ph₃P (19.7 g, 75.1 mmol), and 4-nitrobenzoic acid (11.2 g, 67.2 mmol) at room temperature in benzene (300 mL), and the resulting orange solution was stirred for 20 h at room temperature. The solution was concentrated under reduced pressure to give a viscous orange oil that was dissolved in a minimal amount of $CH_2Cl_2$ and purified twice by flash chromatography; the first column was eluted with 5% ether/hexanes to give the partially purified ester which was resubjected to chromatography eluting with 2% ether/hexanes to give 6.80 g (80%) of 4-nitrobenzoate ester as a white crystalline solid.

The nitrobenzoate ester (6.80 g, 12.0 mmol) in a mixture of THF/MeOH (10 mL:300 mL), containing powdered sodium hydroxide (1.56 g, 39.0 mmol), was stirred at room temperature for 15 min. The reaction was concentrated under reduced pressure, and the residue was partitioned in ether/water (100 mL, 1:1 by volume). The layers were separated, and the aqueous portion was extracted with ether (3 × 50 mL). The combined organic extracts were washed with brine, dried ($MgSO_4$), filtered, and concentrated under reduced pressure. The crude residue was purified by flash chromatography eluting with 5% ether/hexanes to provide 5.00 g (99%) of the inverted alcohol as a clear oil.

Reference: Martin, S. F.; Dodge, J. A.; Burgess, L. E.; Limberakis, C.; Hartmann, M. *Tetrahedron* **1996**, *52*, 3229–3246.

### 2.1.1.8.2 Formation of a C–N Bond

A suspension of triphenylphosphine (590 mg, 2.25 mmol) and 6-chloropurine (348 mg, 2.25 mmol) in anhydrous THF (10 mL) was treated with DEAD (355 μL, 2.25 mmol) at room temperature in the dark for 1 h. A solution of the alcohol (140 mg, 0.56 mmol) in anhydrous THF was then added and the mixture was stirred in the dark at room temperature for 6 h. Evaporation of solvent gave a crude product which was purified by flash chromatography (4:1 hexanes/EtOAc) to yield 149 mg (69%) of the product as a white solid.

Reference: Chong, Y.; Gumina, G.; Chu, C. K. *Tetrahedron: Asymmetry* **2000**, *11*, 4853–4875.

### 2.1.2 Amines

#### 2.1.2.1 Mannich Reaction

The Mannich reaction is an aldol reaction with an imine. It has been the subject of many excellent reviews. For example, see Arend, M.; Westerman, B.; Risch, N. *Angew. Chem. Int. Ed.* **1998**, *37*, 1044–1070. Asymmetric variants have also been reported; see, for example, Notz, W.; Tanaka, F.; Barbas, C. F., III; *Acc. Chem. Res.* **2004**, *37*, 580–591.

##### 2.1.2.1.1 Traditional

Dimethylamine hydrochloride (1 equivalent) was prepared by evaporation of a mixture of aqueous dimethylamine (45 g, 1.0 mol, as 25% aqueous solution) and excess concentrated hydrochloric acid under reduced pressure. To the solid residue was added an aqueous solution of the cyclohexanone (224 g, 2.0 mol) and formaldehyde (30 g, 1.0 mol, 40% aqueous solution). The two-phase mixture was heated carefully (reaction is exothermic) to boiling under a long reflux condenser, boiled for about 5 min, and then cooled to room temperature. Water (200 mL) was added, the layers separated, the aqueous layer saturated with sodium chloride, washed with ether (4 × 50 mL), and then made basic with 30% aqueous potassium hydroxide (1.3 equivalents). The Mannich base separated as a yellow upper layer with a strong amine odor. The layers were separated, the aqueous layer extracted with ether (5 × 100 mL), the combined organic layer and ether extracts dried over magnesium sulfate, and the solvent was distilled to provide the amino ketone (118 g, 70%).

Reference: Frank, R. L.; Pierle, R. C. *J. Am. Chem. Soc.* **1951**, *73*, 724–730.

##### 2.1.2.1.2 Vinylogous Mannich

Reviews: (a) Bur, S. K.; Martin, S. F. *Tetrahedron* **2001**, *57*, 3221–3242. (b) Casiraghi, G.; Zanardi, F.; Appendino, G.; Rassu, G. *Chem. Rev.* **2000**, *100*, 1929–1972.

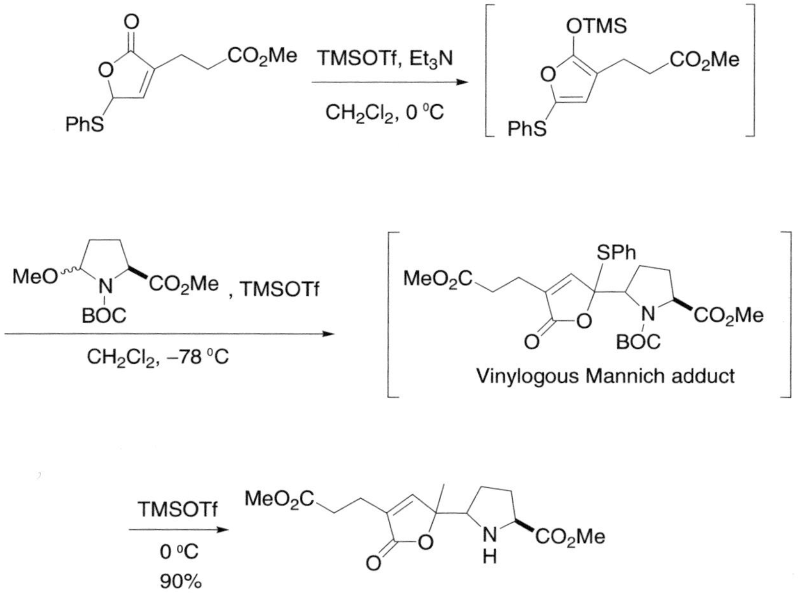

Freshly distilled trimethlysilyl triflate (TMSOTf 461 μL, 2.55 mmol) was added to a mixture of the butenolide (644 mg, 2.31 mmol) and triethylamine (387 μL, 2.78 mmol) in $CH_2Cl_2$ (23 mL) at 0 °C. (In situ formation of the siloxyfuran.) The mixture was stirred at 0 °C for 1 h and then cooled to −78 °C. A solution of the aminal (660 mg, 2.55 mmol) in $CH_2Cl_2$ (3 mL) and then TMSOTf (84 μL, 0.463 mmol) were added, and the reaction mixture was stirred at −78 °C for 1 h. (Formation of the vinylogous Mannich adduct.) Another quantity of TMSOTf (838 μL, 4.63 mmol) was then added. The cooling bath was exchanged for an ice-bath, and the mixture was stirred at 0 °C for 2 h. (BOC deprotection) The mixture was poured into saturated aqueous sodium bicarbonate (30 mL) and the layers were separated. The aqueous layer was extracted with EtOAc (2 x 30 mL), and the combined organic layers were dried ($MgSO_4$), filtered, and concentrated in vacuo. The resulting yellow oil was purified by flash chromatography eluting with EtOAc/hexanes (2:1) to give 839 mg (90%) of the diastereomeric amino butenolides.

Reference: Reichelt, A.; Bur, S. K.; Martin, S. F. *Tetrahedron* **2002**, *58*, 6323–6328.

### 2.1.3 Halides

#### 2.1.3.1 Hartwig-Buchwald Aromatic Amination

Although this reaction is relatively new, it has already been the subject of several reviews: Wolfe, J. P.; Wagaw, S.; Marcoux, J.-F.; Buchwald, S. L. *Acc. Chem. Res.* **1998**, *31*, 805–818; Hartwig, J. F. *Acc. Chem. Res.* **1998**, *31*, 852–860; and Hartwig, J. F. *Angew. Chem. Int. Ed.* **1998**, *37*, 2046–2067. Innumerable conditions have been developed; however, probably the easiest conditions to try for a first pass reaction are shown below.

## FUNCTIONAL GROUP MANIPULATIONS

An oven-dried, resealable Schlenk flask was charged with the palladium source (2.3 mg, 0.005 mmol) and NaO*t*-Bu (134 mg, 1.4 mmol). The flask was evacuated, backfilled with argon, and the aryl chloride (126 mg, 1 mmol), toluene (1 mL), and the amine (105 mg, 1.2 mmol) were then added. The flask was sealed with a Teflon screwcap and the mixture was stirred at 80 °C. After all starting material had been consumed, as judged by gas chromatography (GC), the mixture was allowed to cool to room temperature and was then diluted with ether (40 mL). The resulting suspension was transferred to a separatory funnel and washed with water (10 mL). The organic layer was separated, dried with $MgSO_4$, and concentrated under vacuum. The crude material was purified by flash chromatography to provide the aniline (172 mg, 97%).

Reference: Zim, D.; Buchwald, S. L. *Org. Lett.* **2003**, *5*, 2413–2415.

### 2.1.3.2 Hartwig-Buchwald Ether Formation

Organometallic conditions to form diaryl ethers and aryl alkyl ethers have been developed. For reviews, see Hartwig, J.F. *Angew. Chem. Int. Ed.* **1998**, *37*, 2046–2067; Muci, A.R.; Buchwald, S.L. Practical Palladium Catalysts for C–N and C–O Bond Formation. In *Topics in Current Chemistry;* Miyaura, N., Ed.; Springer-Verlag: Berlin, **2001**, Vol. *219*, pp. 131–209.

A 4-mL vial was charged with bromobenzene (63 mg, 0.40 mmol), Pd(dba)$_2$ (11.5 mg, 0.0200 mmol), Ph$_5$FcP-(*t*-Bu)$_2$ (14.2 mg, 0.0200 mmol), and sodium *tert*-butoxide (47 mg, 0.48 mmol). Anhydrous toluene (2 mL) was added, and the vial was sealed with a cap containing a PTFE septum and removed from the dry box. The reaction mixture was stirred at room temperature for 23 h. The reaction solution was then adsorbed onto silica gel, and the product was isolated by eluting with EtOAc/hexanes (0 to 10% gradient) to give the ether (58 mg, 97%).

Reference: Kataoka, N.; Shelby, Q.; Stambuli, J. P.; Hartwig, J. F. *J. Org. Chem.* **2002**, *67*, 5553–5566.

### 2.1.3.3 Finkelstein Reaction

The chloride (7.20 g, 36.64 mmol) and Aliquat 336 (0.7 g, 1.73 mmol) were added to aqueous LiBr (6.38 g, 73.47 mmol). The mixture was heated to 60 °C for 2 h. After cooling, the mixture was filtered through Florisil to yield the bromide as a pale yellow oil (8.53 g, 97%).

Reference: Chen, B.; Ko, R. Y. Y.; Yuen, M. S. M.; Cheng, K.-F.; Chiu, P. *J. Org. Chem.* **2003**, *68*, 4195–4205.

### 2.1.3.4 Nucleophilic Aromatic Substitution ($S_NAr$)

A variety of electron withdrawing groups activate aromatic rings towards nucleophilic aromatic substitution, including nitro, carboxyl, and cyano groups. The leaving group can be ortho or para to the electron withdrawing group. Fluoride is the most common leaving group, but other halogens and sometimes sulfonate esters are also used.

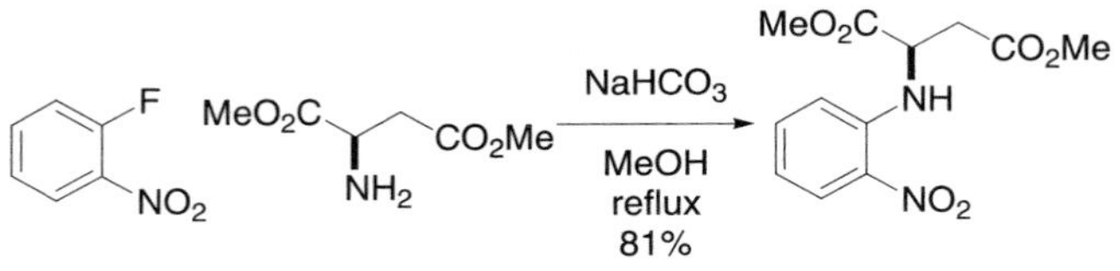

[reaction scheme: 2-fluoronitrobenzene + MeO$_2$C-CH(NH$_2$)-CH$_2$-CO$_2$Me → (via NaHCO$_3$, MeOH, reflux, 81%) 2-nitrophenyl-NH-CH(CO$_2$Me)-CH$_2$-CO$_2$Me]

To a solution of dimethyl *D*-aspartate (18 g, 111.7 mmol) in MeOH (500 mL) was added 2-fluoronitrobenzene (17.33 g, 122.86 mmol) and NaHCO$_3$ (9.38 g, 111.7 mmol). The reaction mixture was refluxed under N$_2$ for approximately 2 days. The solvent was removed under reduced pressure, and the residue was azeotropically dried with benzene (2 × 100 mL). The crude material was then redissolved in MeOH (200 mL), cooled to 0 °C, and the pH of the reaction mixture was adjusted to 4 with HCl(g). The reaction mixture was stirred overnight at room temperature and concentrated under reduced pressure. The residue was taken up in EtOAc and washed with a saturated NaHCO$_3$/10% Na$_2$CO$_3$ solution (9:1, 2 × 500 mL) and brine (300 mL). The organic layer was dried over sodium sulfate, filtered, and concentrated to give dimethyl (2*R*)-2-(2-nitrophenylamino)butanedioate (25.54 g, 81%).

Reference: Su, D.-S.; Markowitz, M. K.; DiPardo, R. M.; Murphy, K. L.; Harrell, C. M.; O'Malley, S. S.; Ransom, R. W.; Chang, R. S. L.; Ha, S.; Hess, F. J.; Pettibone, D. J.; Mason, G. S.; Boyce, S.; Freidinger, R. M.; Bock, M. G. *J. Am. Chem. Soc.* **2003**, *125*, 7516–7517.

## 2.1.4 Olefins

### 2.1.4.1 Hydroboration Reaction

There are a wide variety of hydroborating reagents, including BH$_3$ complexes, pinacolborane, thexyl borane, and catechol borane. The following employs 9-borabicyclo[3.3.1]nonane (9-BBN), which places the boron on the less sterically hindered carbon with high regioselectivity; however, completely removing the cyclooctane by-products can be problematic. The alkylborane can be isolated, but is typically used directly in the next reaction, in this case oxidation to the primary alcohol.

Into a 100-mL round bottomed flask equipped with a nitrogen inlet and a magnetic stir bar was placed the triene (1.09 g, 3.22 mmol) followed by addition of 9-BBN (0.43 M solution in THF, 8.22 mL, 3.54 mmol) via a syringe. After 1 h 20 min of stirring at room temperature, water (8.0 mL) was added, followed by $NaBO_3 \cdot 4H_2O$ (2.48 g, 16.1 mmol). Moderate heat evolution was observed. The heterogeneous mixture was vigorously stirred for 1 h 10 min, quenched with a saturated aqueous solution of $NH_4Cl$ (30 mL), and extracted with a mixture of 1:1 hexane/MTBE (3 × 50 mL). The combined organic layers were dried ($CaCl_2$), filtered, and concentrated. The colorless oil was purified by chromatography (silica gel (20 g), hexane/EtOAc, gradient 10/1 to 4/1, 20 mm column) to afford the alcohol as a colorless oily material (911 mg, 79%).

Reference: Denmark, S. E.; Baiazitov, R. Y. *Org. Lett.* **2005**, *7*, 5617–5620.

### 2.1.4.2 Michael Addition

A variety of nucleophiles can be added in a conjugate manner to α,β-unsaturated ketones. The reaction is reversible, so the main difficulty is finding conditions that drive the equilibrium to the right. For catalytic enantioselective Michael reactions, see Krause, N.; Hoffmann-Röder, A. *Synthesis* **2001**, 171–196. For intramolecular Michael reactions see: Little, R.D.; Masjedizadeh, M. R.; Wallquist, O.; Mcloughlin, J. I. *Org. React.* **1995**, *47*, 315–552.

2-Cyclohexenone (9.68 mL, 0.1 mol) and the amine (15.92 g, 0.1 mol) were combined with $H_2O$ (2.5 mL). Within minutes the mixture became warm and turned solid. This solid was dissolved in EtOH (250 mL), and the solution was heated to reflux for 4 h. The reaction mixture was cooled, dried over $MgSO_4$, and concentrated to give the amino ketone as an unstable yellow solid (24.0 g, 93%).

Reference: Wright, J. L.; Caprathe, B. W.; Downing, D. M.; Glase, S. A.; Heffner, T. G.; Jaen, J. C.; Johnson, S. J.; Kesten, S. R.; MacKenzie, R. G.; Meltzer, L. T.; Pugsley, T. A.; Smith, S. J.; Wise, L. D.; Wustrow, D. J. *J. Med. Chem.* **1994**, *37*, 3523–3533.

## 2.2 Ketone Oxidation State

### 2.2.1 Alkyne Hydrolysis

The hydrolysis has traditionally been performed with protic acids. Recently, however, transition metals have been used to facilitate the reaction. Frequently, with internal alkynes, a mixture of regioisomeric ketones are formed.

1-Octyne (7.5 g) and formic acid (100 mL) were heated in an oil bath at 100 °C until all starting material was consumed. The progress of the reaction was monitored by GC analysis of the reaction solution. Quantitative GC analysis at the end of the reaction (6 h) indicated 92% yield of 2-octanone. The cooled reaction mixture was taken up with $CH_2Cl_2$ (170 mL), and the solution was washed with water, sodium carbonate solution, and water, dried over $MgSO_4$, and evaporated in vacuo. The residue was distilled (bp 171–173 °C) to give 2-octanone (7.42 g, 85%).

Reference: Menashe, N.; Reshef, D.; Shvo, Y. *J. Org. Chem.* **1991**, *56*, 2912–2914.

### 2.2.2 Epoxides

#### 2.2.2.1 Epoxide Ring Opening Reactions, $S_N1$

Note: *Perchloric acid is an extremely corrosive acid. The proper personal protection equipment must be worn. In addition, perchloric acid must be handled in a hood, preferably in a hood designed for it. Avoid contact with paper, cotton, and wood as these materials may catch fire. If perchloric acid becomes concentrated, a fire or explosion could result. Read its Material Safety Data Sheet for more details.*
A suspension of the epoxide (144 mg, 0.36 mmol) in THF (2.5 mL) and $H_2O$ (1.4 mL) at room temperature was added to $HClO_4$ (70% in $H_2O$, 20 μL, 0.23 mmol). After 50 min at room temperature, $NaHCO_3$ (20 mL, saturated) was added and the aqueous layer was extracted with ether (4 × 20 mL). The combined organic extracts were washed with saturated $NaHCO_3$ (30 mL) and brine (30 mL), dried ($Na_2SO_4$), and evaporated to give a pale yellow oil (160 mg). The crude product was purified by flash chromatography ($SiO_2$, 40% EtOAc in hexanes) to give the diol as a colorless oil (124 mg, 82%).

Reference: Jin, Q.; Williams, D. C.; Hezari, M.; Croteau, R.; Coates, R. M. *J. Org. Chem.* **2005**, *70*, 4667–4675.

#### 2.2.2.2 Epoxide Ring Opening Reactions, $S_N2$

To a stirred solution of the epoxide (3.50 g, 40.6 mmol) and CuCN (364 mg, 4.06 mmol) in dry THF (30 mL) was added a 1 M solution of vinylmagnesium bromide in THF (52.8 mL, 52.8 mmol), over 45 min, dropwise at −78 °C. The mixture was allowed to warm up to 0 °C before it was quenched with a saturated aqueous NH$_4$Cl solution (20 mL). The layers were separated, the aqueous layer extracted with ether (3 × 50 mL), and the combined ethereal extracts were washed with brine (20 mL) and dried (MgSO$_4$). Evaporation of the solvent and chromatographic purification of the crude product (silica, Et$_2$O:pentane 1:3) gave the alcohol as a pale yellow oil (4.41 g, 95%).

Reference: Holub, N.; Neidhoefer, J.; Blechert, S. *Org. Lett.* **2005**, *7*, 1227–1229.

### 2.2.3 Ketones

#### 2.2.3.1 Ketal Formation

The most common ketals are dioxolane and dioxane; these form more readily and have fewer problems with enol ether contamination. Sometimes, however, they can be difficult to hydrolyze, particularly when the molecule contains sensitive functional groups. The reaction is acid catalyzed, with some of the more common acids being para-toluenesulfonic, pyridinium para-toluenesulfonic, and camphorsulfonic. The eliminated water can be removed either with an azeotroping solvent (such as toluene or benzene) or by the addition of a dehydrating agent such as trimethylorthoformate.

A solution of the ketone (1.23 g, 8.11 mmol), ethylene glycol (1.8 mL, 32 mmol), pyridinium tosylate (0.6 g, 2.4 mmol), and benzene (45 mL) was refluxed for 22 h in a Dean–Stark apparatus. The reaction mixture was cooled, poured into saturated NaHCO$_3$ (50 mL), the aqueous layer extracted with hexanes/Et$_2$O (1/1, 2 × 20 mL), and washed with brine (2 × 15 mL). The combined organics were dried (MgSO$_4$), concentrated, and chromatographed to give the ketal (1.59 g, 100%).

Reference: Behenna, D. C.; Stoltz, B. M. *J. Am. Chem. Soc.* **2004**, *126*, 15044–15045.

#### 2.2.3.2 Enamine Formation

As with imines, enamines have varying stability. They are typically used crude without further purification.

A 250-mL round-bottomed flask equipped with a Dean–Stark trap and reflux condenser was charged with the ketone (5.2 g, 33.3 mmol) and pyrrolidine (2.4 mL, 33.3 mmol) in anhydrous benzene (200 mL). The solution was refluxed for 8 h until no more $H_2O$ was collected in the Dean–Stark trap. The Dean–Stark trap was removed and replaced with a distillation head, and the benzene was removed by distillation. The product was distilled (0.2 torr, 105 °C) to give the enamine (4.78 g, 69%) as a yellow viscous oil, which was used immediately to avoid decomposition.

Reference: Davis, K. M.; Carpenter, B. K. *J. Org. Chem.* **1996**, *61*, 4617–4622.

### 2.2.3.3 Imine Formation

Imines display a variety of stabilities—some condensations of anilines with benzaldehydes liberate water before any heat can be applied while some imines cannot survive TLC and, as in this case, must be stored under special conditions to prevent hydrolysis.

4-Pentynyl-1-amine (0.59 g, 6.7 mmol) was dissolved in dry benzene (15 mL), and 4 Å molecular sieves (~ 4 g) were added to the solution. The mixture was stirred at room temperature and acetophenone (0.78 mL, 6.7 mmol) was added dropwise. The mixture was stirred until $^1$H NMR spectroscopy revealed complete imine formation (~ 6 h). The reaction mixture was filtered through a plug of dry Celite, and the molecular sieves were washed thoroughly with ether. The solution was concentrated in vacuo to yield the imine (1.2 g, 94%) as a clear oil, which was used without further purification. The imine can be stored frozen in benzene to avoid hydrolysis.

Reference: Prabhakaran, E. N.; Nugent, B. M.; Williams, A. L.; Nailor, K. E.; Johnston, J. *J. Org. Lett.* **2002**, *4*, 4197–4200.

### 2.2.3.4 Oxime Formation

Oximes are formed in much the same way as imines (condensation with a carbonyl, typically driven by removing the formed water). Oximes are usually much more stable than imines and can be more easily purified.

To a solution of the keto-ester (1.44 g, 10 mmol, 1.0) and benzyl-hydroxylamine (1.60 g, 10 mmol, 1.0) in ethanol (30 mL) was added pyridine (5.0 mL, 62 mmol) in 1 portion. The reaction mixture was heated at 55 °C for 24 h and then concentrated on a rotary evaporator. The residue was partitioned between ether (150 mL) and water (50 mL). The organic layer was sequentially washed with HCl (0.5 N, 2 × 30 mL) and water (30 mL), and then dried over MgSO$_4$. Concentration in vacuo provided the oxime (2.50 g, 100%) as a 3:1 mixture of E:Z isomers as a colorless liquid.

Reference: Hart, D. J.; Magomedov, N. A. *J. Am. Chem. Soc.* **2001**, *123*, 5892–5899.

### 2.2.3.5 Sulfinimine Formation

One of the most effective ways of preparing enantiomerically pure secondary amines is the addition of organometallic reagents to chiral sulfinimines, which are prepared by the condensation of an aldehyde (or ketone) with a sulfinamide. The preparation of *t*-butyl sulfinamide had been problematic, but the synthesis (and supplies) now seems to be more reliable.

A mixture of *t*-butyl sulfinamide (1.52 g, 12.5 mmol), the ketone (2.37 g, 15.0 mmol), and Ti(OEt)$_4$ (6.0 g, 25.0 mmol) in 25 mL of THF was heated to reflux for 15 h and then was cooled to room temperature. The reaction mixture was poured into 25 mL of brine while rapidly stirring, and the resulting mixture was filtered through a plug of Celite. The filter cake was washed with EtOAc, and the combined filtrate was transferred to a separatory funnel where the organic layer was washed with brine. The brine layer was extracted once with EtOAc, and the combined organic portions were dried over Na$_2$SO$_4$, filtered, and concentrated. Column chromatography (4:6 hexanes/EtOAc) afforded the sulfinimine (3.21 g, 82%) as a mixture of E/Z isomers (8:1 by $^1$H NMR in CDCl$_3$).

Reference: Kochi, T.; Tang, T. P.; Ellman J. A. *J. Am. Chem. Soc.* **2003**, *125*, 11276–11282.

### 2.2.3.6 Thioketone Formation

The two primary reagents for accomplishing this transformation are P$_4$S$_{10}$ (alternately referred to as P$_2$S$_5$) and Lawesson's reagent. The primary advantage of P$_4$S$_{10}$ is that it is far less expensive and more atom economical. The primary advantage of Lawesson's reagent is that it can be used with only a moderate excess, whereas P$_4$S$_{10}$ requires a large excess and is easier to handle. Both of these reagents have the stench one might expect from a compound containing multiple sulfur atoms.

The ketone (1.00 g, 3.17 mmol) and 2,4-bis(4-methoxyphenyl)-1,3-dithia-2,4-diphosphetane 2,4-disulfide (Lawesson's reagent, 3.85 g, 9.52 mmol) were mixed in PhMe (80 mL) and heated at reflux for 12 h. The mixture was allowed to cool to room temperature, filtered through a plug of silica, and concentrated in vacuo. Chromatography (1:3 EtOAc:hexane; $R_f$ 0.27) followed by recrystallization (hexane) gave the thione (0.86 g, 82%) as orange crystals.

Reference: Barrett, A. G. M.; Braddock, D. C.; Christian, P. W. N.; Pilipauskas, D.; White, A. J. P.; Williams, D. J. *J. Org. Chem.* **1998**, *63*, 5818–5823.

## 2.3 Acid Oxidation State

### 2.3.1 Reaction by Reactive Intermediates

Because the intermediates described in this section are reactive and potentially unstable compounds, the procedures describe the preparation of the active intermediate followed by a subsequent reaction rather than just the active intermediate.

#### 2.3.1.1 Acid Chloride Formation by Oxalyl Chloride

One of the most straightforward methods of activating acids via the acid chloride is by reacting the acid with oxalyl chloride and a catalytic amount of DMF. This reaction is fast and liberates a great deal of gas, so it is important that the reaction vessel is adequately vented; it is also moderately exothermic and frequently requires cooling. As the intermediate acid chloride typically hydrolyzes on silica gel, the reaction can be monitored by first dropping an aliquot of the reaction mixture into methanol and then checking by TLC for disappearance of the acid and formation of the methyl ester.

To a room temperature solution of Fmoc-*L*-alanine (1.04 g, 3.35 mmol) in dichloromethane (5 mL) was added DMF (26 µL) followed by oxalyl chloride (584 µL, 6.69 mmol). Once the effervescence had subsided, the yellow solution was heated to reflux for 1.5 h. After cooling, the solution was evaporated in vacuo and the solid residue azeotroped with benzene, before being dissolved in benzene (6 mL). To this yellow solution was added a solution of *N-tert*-butyl-*O*-benzoylhydroxylamine (0.71 g, 3.68 mmol) in benzene (5 mL), followed by pyridine (541 µL, 6.69 mmol). The mixture was heated to reflux overnight and then cooled. The colorless suspension was taken up in EtOAc and 10% HCl (aqueous), separated, the organic layer washed with brine, and then dried over $MgSO_4$. Evaporation gave a viscous, colorless oil that was purified by flash chromatography (4:1 Hexane:EtOAc) to give the hydroxamic acid as a colorless oil (1.56 g, 96%).

Reference: Braslau, R.; Axon, J. R.; Lee, B. *Org. Lett.* **2000**, *2*, 1399–1401.

## 2.3.1.2 Acid Chloride Formation by Phosphorous Oxychloride

An advantage of phosphorous oxychloride in the preparation of acid chlorides is that no gas is evolved. As is demonstrated in the following example, basic nitrogens do not interfere with the reaction.

To a cooled solution (0 °C) of the amino acid (193 mg, 1 mmol) and triethylamine (418 µL, 3 mmol) in $CH_2Cl_2$ (25 mL) was added $POCl_3$ (460 mg, 3 mmol) dropwise, and the reaction mixture was stirred overnight at room temperature. The resulting solution was washed with saturated aqueous $NaHCO_3$ (25 mL), brine (25 mL), and water (3 × 25 mL). Drying over anhydrous $Na_2SO_4$ and evaporation of the solvent yielded the β-lactam (145 mg, 83%).

Reference: Sharma, S. D.; Anand, R. D.; Kaur, G. *Synth. Commun.* **2004**, *34*, 1855–1862.

## 2.3.1.3 Acid Chloride Formation by Thionyl Chloride

Thionyl chloride costs less and liberates less gas than oxalyl chloride.

Thionyl chloride (10.4 g, 87.4 mmol) was added to a mixture of biphenyl-2-carboxylic acid (15.0 g, 75.7 mmol) and DMF (0.28 g, 3.83 mmol) in toluene (72 mL) at an internal temperature of 40 °C. The mixture was stirred at this temperature for approximately 2 h. After completion of the reaction, the mixture was concentrated to dryness at 60 °C. The resultant residue was then diluted with toluene (36 mL) and concentrated to dryness at 60 °C, and the process was repeated again to give biphenyl-2-carbonyl chloride as an oil. Acetone (100 mL) was added to the oil, and 4-amino benzoic acid (10.4 g, 75.8 mmol) and *N,N*-dimethylaniline (10.1 g, 83.3 mmol) were added to the resultant solution at 25 °C. The mixture was stirred at this temperature for approximately 2 h. Water (100 mL) was then poured into the mixture, and it was stirred at 25 °C for more than 1 h. The resultant crystals were collected by filtration and dissolved in DMF (100 mL) at 25 °C. The solution was then filtered to remove insoluble materials, water (100 mL) was poured into the filtrate, and it was stirred at 25 °C for approximately 2 h. The resultant crystals were collected by filtration and dried at 40 °C to give the amide as white crystals (22.7 g, 95%).

Reference: Tsunoda, T.; Yamazaki, A.; Mase, T.; Sakamoto, S. *Org. Process Res. Dev.* **2005**, *9*, 593–598.

### 2.3.1.4 Carbonyldiimidazole Activation

Acyl imidazolides can be used as activated acids; they are very frequently used to prepare β-keto esters from acids. It has recently been found that carbon dioxide dramatically increases the rate of reaction of the acyl imidazolide with an amine (see Vaidyanathan, R.; Kalthod, V. G.; Ngo, D. P.; Manley, J. M.; Lapekas, S. P.; *J. Org. Chem.* **2004**, *69*, 2565–2568).

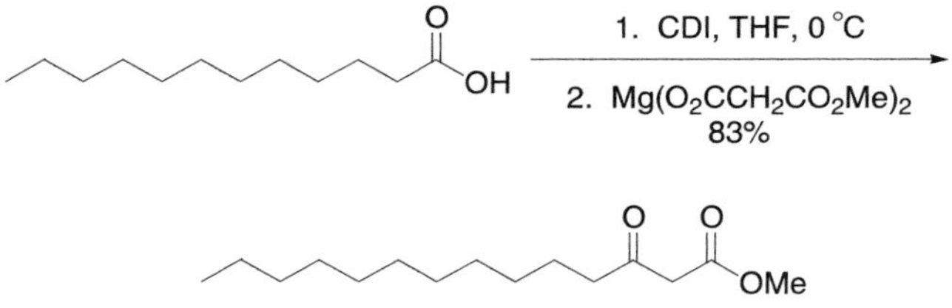

$$\text{R-CO-OH} \xrightarrow[\text{2. Mg(O}_2\text{CCH}_2\text{CO}_2\text{Me})_2]{\text{1. CDI, THF, 0 °C}} \text{R-CO-CH}_2\text{-CO-OMe}$$
83%

The carboxylic acid (13.8 g, 69 mmol, 1.0 equivalents) was dissolved in dry THF under an inert atmosphere and cooled to 0 °C. CDI (13.4 g, 83 mmol, 1.2 equivalents) was then added in small portions over several minutes. After 10 min, the reaction was allowed to warm slowly to room temperature and was then stirred for 1 h. In a separate flask, monomethyl malonate (9.8g, 83 mmol, 1.2 equivalents) was dissolved in THF under an inert atmosphere and cooled to −78 °C. To this solution was added dibutylmagnesium (1.0 M in heptane, 0.6 equivalents). A white solid formed instantly on addition of the base. After 10 min, the reaction was warmed to room temperature and stirred for 1 h. The acylimidazole was then added by cannula to the flask containing the magnesium salt. The resulting slurry was stirred for 3 days. The reaction mixture was then concentrated on a rotary evaporator and the residue redissolved in EtOAc. The resulting solution was washed with 1.2 M HCl, saturated aqueous NaHCO$_3$, and brine. The organic phase was dried over sodium sulfate, filtered, concentrated, and the residue purified by flash chromatography (7:1 hexanes:EtOAc) to give the β-keto ester (14.7g, 83%).

Reference: Durham, T. B.; Miller, M. J. *J. Org. Chem.* **2003**, *68*, 27–34.

### 2.3.1.5 EDCI Activation

1-Ethyl-3-[3-(dimethylamino)propyl]carbodiimide hydrochloride (EDCI), which is frequently used in amino acid couplings, is very mild and results in minimal epimerization. EDCI is shown here as representative of all carbodiimide coupling reagents. The advantage of EDCI is that the urea by-product can be extracted into an aqueous acidic layer, whereas many times complete removal of the urea from the final product can be problematic.

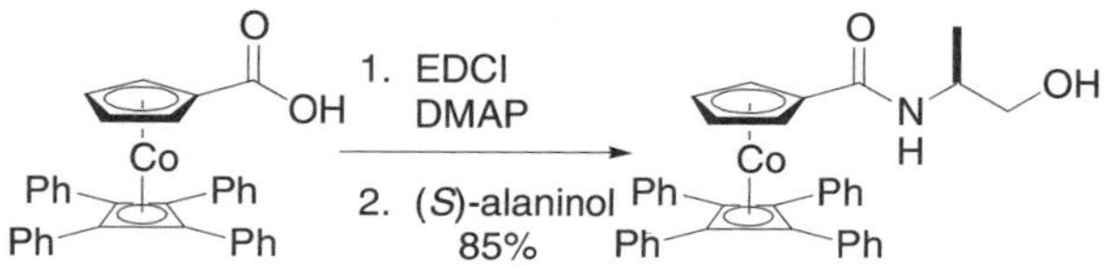

To a solution of the acid (2.00 g, 3.81 mmol) in $CH_2Cl_2$ (30 mL) under a nitrogen atmosphere were added EDCI (0.880 g, 4.59 mmol), *N,N*-dimethylaminopyridine (DMAP 558 mg, 4.57 mmol), and a solution of (*S*)-alaninol (0.343 g, 4.57 mmol) in $CH_2Cl_2$ (20 mL), and the resulting mixture stirred for 48 h at room temperature. The reaction was quenched by the addition of 10% aqueous citric acid (20 mL), followed by stirring for 5 min. The mixture was then partitioned and the orange organic fraction dried ($Na_2SO_4$) and evaporated in vacuo to afford a solid, which was purified by column chromatography (EtOAc) to give the amide as an orange crystalline solid (1.88 g, 85%).

Reference: Prasad, R. S.; Anderson, C. E.; Richards, C. J.; Overman, L. E. *Organometallics* **2005**, *24*, 77–81.

### 2.3.1.6 EEDQ Activation

2-Ethoxy-1-ethoxycarbonyl-1,2-dihydroquinoline (EEDQ) activates acids via a mixed anhydride and generates quinoline, ethanol, and carbon dioxide as by-products. As with EDCI, the by-product quinoline of the reaction can be extracted into an aqueous acidic medium.

To a solution of *N*-BOC-*L*-proline (85 g, 0.40 mol) and EEDQ (100 g, 0.41 mol) in $CH_2Cl_2$ (200 mL) was added 5-aminoindan (54 g, 0.41 mol) at 0 °C. The reaction mixture was stirred at 0 °C for 2 h and then allowed to react overnight at room temperature. The reaction mixture was diluted with 100 mL of $CH_2Cl_2$ and washed with 1 M HCl (3 × 50 mL), saturated aqueous $NaHCO_3$ (2 × 50 mL), $H_2O$ (50 mL), and brine (2 × 50 mL). The organic phase was dried over $MgSO_4$, filtered, and concentrated to give a viscous oil. The crude product was then redissolved in a 1:1 mixture of $CH_2Cl_2$/EtOAc. Cooling the mixture to –10 °C afforded the amide (116 g, 90%) as an off-white crystalline solid.

Reference: Ling, F. H.; Lu, V.; Svec, F.; Frechet, J. M. J. *J. Org. Chem.* **2002**, *67*, 1993–2002.

### 2.3.1.7 Isobutylchloroformate Activation

The mixed anhydride formed by the reaction of a carboxylic acid with isobutyl chloroformate is frequently more stable than the corresponding acid chloride, so epimerization of an α-stereogenic center is minimized.

In a 1-L, four-neckedflask equipped with an addition funnel, a low temperature thermometer, an $N_2$ inlet, and a mechanical stirrer, N-BOC-(3-fluorophenyl)alanine (56.6 g, 0.2 mol) was dissolved in $CH_2Cl_2$ (300 mL). N-Methylmorpholine (23.05 mL, 0.2 mol) was added in a slow stream with a slight exotherm of 18–24 °C. The solution was cooled to –25 °C, and isobutyl chloroformate (25.27 mL, 0.2 mol) was added over 2–3 min while keeping the temperature between –25 and –20 °C. A precipitate formed as the reaction mixture was stirred at –20 to –10 °C for 1 h. In a separate flask, a slurry of N,O-dimethylhydroxylamine hydrochloride (21.45 g, 0.22 mol) in $CH_2Cl_2$ (200 mL) was treated with N-methylmorpholine (24.15 mL, 0.22 mol) at room temperature. The reaction remained a slurry throughout as N-methylmorpholine hydrochloride formed. After 1 h, the hydroxylamine suspension was added over 30 min to the mixed anhydride with the temperature rising to 5 °C. The mixture was stirred at room temperature over the weekend. (The reaction was probably done on addition.) The reaction was quenched by the addition of a solution of citric acid (50 g) in water (200 mL). The organic layer was separated and washed with water, saturated aqueous $NaHCO_3$, and brine. The organic layer was dried over $MgSO_4$, filtered, and evaporated to an oil which was dried under high vacuum to remove residual solvents to provide the hydroxamate ester (61.7 g, 97%).

Reference: Urban, F. J.; Jasys, V. *J. Org. Process Res. Dev.* **2004**, *8*, 169–175.

## 2.3.1.8 Mukaiyama Esterification

A variety of alkyl pyridinium salts have been used in the Mukaiyama coupling; a representative example is shown below.

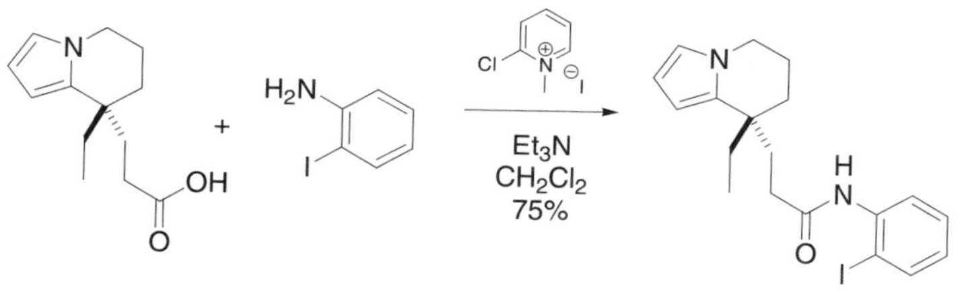

To a solution of the acid (300 mg, 1.1 mmol) in $CH_2Cl_2$ (5 mL) was added 2-chloro-1-methyl-pyridinium iodide (the Mukaiyama reagent, 330 mg, 1.3 mmol) and the aniline (1.29 g, 5.89 mmol). The mixture was heated at reflux for 1 h and then allowed to cool. $Et_3N$ (0.30 mL, 2.2 mmol) was added, and the mixture was heated at reflux for an additional 20 h, allowed to cool, poured into water (40 mL), and extracted with $CH_2Cl_2$ (3 × 20 mL). The combined organic extracts were dried, filtered, and concentrated. The crude material was purified by column chromatography (25% EtOAc in hexanes) to furnish the amide (387 mg, 75%).

Reference: Bowie, Jr., A. L.; Hughes, C. C.; Trauner, D. *Org. Lett.* **2005**, *7*, 5207–5209.

## 2.3.1.9 Yamada Coupling

Yamada's coupling reagent (diethylcyanophosphonate, DEPC) is representative of phosphorous based activators, including diphenylphosphoryl azide.

A solution of the acid (168 mg, 0.371 mmol) in CH$_2$Cl$_2$ (5 mL) was treated with $i$-Pr$_2$NEt (400 µL, 2.30 mmol), cooled to −30 °C, and then DEPC (100 µL, 0.659 mmol) was added. The reaction was warmed to −20 °C over a 20-min period, a solution of HCl·NH$_2$-$L$-Leu-OBn (282 mg, 1.09 mmol) in CH$_2$Cl$_2$ (1 mL) was added and stirring was continued for another 2 h. The solution was partitioned between Et$_2$O and 10% HCl, and the organic layer was washed with 2 N NaOH, water, dried (MgSO$_4$), filtered, and concentrated. Chromatography on SiO$_2$ (hexanes/EtOAc, 9:1) provided the amide (178 mg, 73%).

Reference: Wipf, P.; Methot, J.-L. *Org. Lett.* **2000**, *2*, 4213–4216.

### 2.3.1.10 Yamaguchi Esterification

The Yamaguchi esterification (activation of an acid with trichlorobenzoyl chloride) is typically used to make macrolides, indicating its reliability. Unfortunately, these examples are typically done on very small scale, so a different example is shown.

To a stirred solution of the acid (100 mg, 248 µmol) in benzene (2.5 mL) were added $i$-Pr$_2$NEt (99.8 µL, 573 µmol), Cl$_3$C$_6$H$_2$COCl (85.6 µL, 548 µmol), and DMAP (151 mg, 1.24 mmol). The alcohol (32.1 mg, 124 µmol) was then added in benzene (1.5 mL). The resulting mixture was diluted with benzene (1.5 mL) and stirring continued for 20 h. Benzene (50 mL) and saturated aqueous NaHCO$_3$ (50 mL) were added. The layers were separated, and the aqueous layer extracted with benzene (2 × 50 mL). The combined organic layers were washed with brine (50 mL), dried over $Na_2SO_4$, and concentrated in vacuo. Flash chromatography (hexane/EtOAc 30:1) yielded the ester (68.0 mg, 85%) as an oil.

Reference: Kangani, C. O.; Brückner, A. M.; Curran, D. P. *Org. Lett.* **2005**, *7*, 379–382.

### 2.3.2 Acid Reaction Without Prior Activation

#### 2.3.2.1 Fischer Esterification

The Fischer esterification is one of the oldest and most reliable methods of converting an acid to an ester. The reaction requires acid catalysis, elevated temperatures, and frequently the removal of the water by-product. As long as the remainder of the starting acid can withstand these requirements, the Fischer esterification deserves consideration.

To methanol (80 mL) cooled to 0 °C (ice/water) was added dropwise concentrated $H_2SO_4$ (25 mL). To the resulting clear solution was added immediately the carboxylic acid (21 g, 112 mmol), and the mixture was refluxed for 3 h, at which time TLC indicated the reaction to be complete. The resulting yellowish emulsion was cooled to room temperature, and the solvent was evaporated as much as possible. The residue was partitioned between $CH_2Cl_2$ and water. The collected organic layer was washed with saturated $NaHCO_3$ (aqueous), collected, dried ($Na_2SO_4$), and decanted. The solvent was concentrated and the resulting brownish oil was subjected to Kugelrohr distillation to afford the ester (20 g, 90%) as a yellowish oil.

Reference: Kolotuchin, S. V.; Meyers, A. I. *J. Org. Chem.* **1999**, *64*, 7921–7928.

#### 2.3.2.2 TMSCHN₂ Esterification

Trimethylsilyldiazomethane is a commercially available equivalent to diazomethane. It reliably converts an acid to the methyl ester. Methanol is an essential co-solvent for this esterification.

To a solution of BOC-Gln(Trt)-OH (10.1 g, 20.8 mmol) in toluene:methanol (7:1, 300 mL) was added a solution of $TMSCHN_2$ (12.5 mL, 2.0 M). The mixture was allowed to stir at room temperature until the evolution of $N_2$ ceased (approximately 6 h). The solvents were removed in vacuo to provide BOC-Gln(Trt)-OMe (10.4 g, 100%).

Reference: Brewer, M.; James, C. A.; Rich, D. H. *Org. Lett.* **2004**, *6*, 4779–4782.

## 2.3.3 Acid Chloride

See also the above section about the reaction of acids, as acid chlorides are frequently not isolated but used immediately after formation.

### 2.3.3.1 Schotten–Baumann Reaction

The Schotten–Baumann reaction is one of the most effective methods of converting acid chlorides to amides or chloroformates to carbamates. The workup typically involves separating the phases, drying the organic layer, and concentrating.

To a solution of 3-fluoroaniline (50.0 g, 450 mmol) in $CH_2Cl_2$ (200 mL) was added a solution of potassium carbonate (46.9 g, 339 mmol) in water (200 mL) at room temperature. The mixture was warmed to 32 °C, and isobutyl chloroformate (66.2 g, 485 mmol) was added over 13 min while maintaining the temperature at 30–35 °C. The mixture was stirred at 30–35 °C for 2.5 h until complete by GC analysis. Aqueous ammonia (29.3 wt%, 7.2 mL, 111 mmol) was added and the mixture stirred at 30–35 °C for 15 min. The mixture was cooled to 25 °C and the pH adjusted from 8.7 to 1.9 with concentrated hydrochloric acid. The phases were separated, and the aqueous layer was washed with $CH_2Cl_2$ (100 mL). The combined organics were washed with water (200 mL), and the water was back-extracted with $CH_2Cl_2$ (100 mL). Typically, the crude product was carried into the next step but, if desired, crystallization at this point from heptane at −30 °C afforded the amide (93.1 g, 98.1%).

Reference: Herrinton, P. M.; Owen, C. E.; Gage, J. R. *Org. Process Res. Dev.* **2001**, *5*, 80–83.

## 2.3.4 Amide

### 2.3.4.1 Dehydration to Nitrile

A variety of dehydrating agents can be used, including $P_2O_5$, TCCA, and tosyl chloride.

To a solution of the amide (1.15 g, 3.6 mmol) and *para*-toluenesulfonyl chloride (1.91 g, 7.2 mmol) in $CH_2Cl_2$ (9 mL) was added pyridine (1.46 mL, 18 mmol). After complete reaction, as judged by TLC, saturated aqueous $NaHCO_3$ was carefully added.

The resultant two-phase mixture was stirred vigorously for 2 h before the layers were separated and the organic phase was washed with 1 M aqueous HCl and then another portion of saturated aqueous $NaHCO_3$. The aqueous phases were back-extracted with $CH_2Cl_2$ and the combined organic phases were dried, filtered, and concentrated to leave a crude product. This crude product was purified by column chromatography to provide the nitrile (1.02g, 94%).

Reference: McLaughlin, M.; Mohareb, R. M.; Rapoport, H. *J. Org. Chem.* **2003**, *68*, 50–54.

### 2.3.4.2 Hydrolysis to Acid

Amides are much more robust than esters but can be hydrolyzed under vigorous conditions.

R-(+)-2-Methyl-2-hydroxy-3-phenylpropionamide (179 mg, 1 mmol) was refluxed in hydrochloric acid (6 *N*, 29 ml) for 3 h to give, after extraction with EtOAc and column chromatography using a silica gel column with a mixture of petroleum ether and EtOAc (1:2) as an eluent, R-(+)-2-methyl-2-hydroxy-3-phenylpropionic acid as a white solid (166 mg, 92%).

Reference: Wang, M.-X.; Deng, G.; Wang, D.-X.; Zheng Q.-Y. *J. Org. Chem.* **2005**, *70*, 2439–2444.

## 2.3.5 Ester Hydrolysis

### 2.3.5.1 Acid Promoted

An acid promoted hydrolysis is usually slower than a similar base promoted hydrolysis. In some cases, as with a *t*-butyl ester, however, acid catalyzed hydrolyses are preferred. With the *t*-butyl ester, sometimes a reducing agent such as formic acid or triethylsilane is added to scavenge the carbocation.

To the solution of the ester (950 mg, 3.31 mmol) in $CH_2Cl_2$ (10 mL) was carefully added an equal volume of $CF_3CO_2H$ (10 mL) through a syringe at room temperature. After being stirred at room temperature for 3 h, the reaction mixture was concentrated in vacuo. The residue was azeotroped with toluene twice to give the acid in a quantitative yield.

Reference: Yang, D.; Qu, J.; Li, W.; Wang, D.-P.; Ren, Y.; Wu Y.-D. *J. Am. Chem. Soc.* **2003**, *125*, 14452–14457.

## 2.3.5.2 Base Promoted

Saponification of an ester with base is rapid and reliable. Lithium, sodium, and potassium hydroxide can be used.

HO~~~OBn~~C(O)OEt  →[LiOH, MeOH:H$_2$O (3:1), 87%]  HO~~~OBn~~C(O)OH

To a cooled (0 °C) solution of the ester (2.33 g, 8.31 mmol) in 32 mL of MeOH/H$_2$O (3:1 v/v) was added LiOH·H$_2$O (0.891 g of hydrate, 56% LiOH, 11.9 mmol of LiOH), and then the solution was warmed to ambient temperature. After 4 h, the reaction mixture was acidified to below pH 2 using 40 mL of 1.0 M HCl; 50 mL of EtOAc and 25 mL of brine were then added to facilitate separation of the layers. The aqueous portion was extracted with 10 mL of EtOAc and the combined organic extracts were washed with 50 mL of brine, dried over anhydrous MgSO$_4$, filtered, and concentrated in vacuo to afford the acid as a colorless oil (1.825 g, 87%).

Reference: Chamberland S.; Woerpel, K. A. *Org. Lett.* **2004**, *6*, 4739–4741.

## 2.3.6 Nitrile

### 2.3.6.1 Acid Promoted Hydrolysis to Acid

Nitrile hydrolysis to amide is relatively facile; subsequent hydrolysis to the acid requires somewhat more forcing conditions.

[Reaction scheme: BnO-substituted acetonide with CN and N-H-S(O)$_2$-tolyl group →(HCl, Et$_2$O/HCl, 95%) BnO~~CH(OH)~CH(OH)~CH(CO$_2$H)(NH$_3$Cl)]

In a 25-mL, single-necked, round-bottomed flask equipped with a magnetic stirring bar was placed 0.21 g (0.5 mmol) of the nitrile. Saturated ethereal HCl (20 mL) was added with vigorous stirring, which resulted in the immediate formation of white precipitate. A few drops of H$_2$O were added, and the reaction mixture was stirred at room temperature for 12 h. At this time the solution was diluted with ether (25 mL), and the precipitated hydrochloride was filtered and dried under vacuum to give the acid as a white solid (0.12g, 95%).

Reference: Davis, F. A.; Prasad, K. R.; Carroll, P. J. *J. Org. Chem.* **2002**, *67*, 7802–7806.

### 2.3.6.2 Acid Promoted Hydrolysis to Amide

F$_3$C-C(OH)(CN)- →[H$_2$SO$_4$, H$_2$O, 48%] F$_3$C-C(OH)-C(O)NH$_2$

2-Hydroxy-2-(trifluoromethyl)butanenitrile (8 g, 52 mmol) was added slowly dropwise to concentrated $H_2SO_4$ (15.3 g). The mixture was heated to 115 °C for 15 min, cooled to 8 °C and 22 g of water added. Diethyl ether (50 mL) was then added, and the organic phase was washed with water (25.0 mL), saturated aqueous $NaHCO_3$ (25.0 mL), and again with water (25.0 mL). The diethyl ether phase was dried over $Na_2SO_4$, filtered, and evaporated. The resulting oil was treated with *n*-hexane, and the resulting amide (4.27 g, 48%) was collected by filtration.

Reference: Shaw, N. M.; Naughton, A. B. *Tetrahedron* **2004**, *60*, 747–752.

### 2.3.6.3 Base Promoted Hydrolysis to Acid

A solution of 5-methylene-6-heptenenitrile (2.0 g, 16.5 mmol) in 25% NaOH (95 mL) and MeOH (300 mL) was heated to reflux for 48 h. The solution was cooled, concentrated to half its original volume, and acidified with concentrated HCl to pH 1. The mixture was extracted with $Et_2O$ (2 × 30 mL), the combined extracts dried ($MgSO_4$), and concentrated in vacuo to afford 2.10 g (90%) of 5-methylene-6-heptenoic acid as a pale yellow oil.

Reference: Sparks, S. M.; Chow, C. P.; Zhu, L.; Shea K. J. *J. Org. Chem.* **2004**, *69*, 3025–3035.

### 2.3.6.4 Base Promoted Hydrolysis to Amide

The base promoted conversion of nitrile to amide is sometimes catalyzed by addition of hydrogen peroxide. Because of safety concerns with peroxides, a procedure without this additive has been selected.

A solution of the nitrile (2.77 g, 14.5 mmol) and powdered 85% KOH (7.66 g, 116 mmol) in *t*-butyl alcohol (30 mL) was heated at reflux for 1.5 h. The reaction mixture was then cooled to room temperature, diluted with water (30 mL), and acidified with 1 N HCl (116 mL, 116 mmol) to give a slurry that was filtered and rinsed with water and then $Et_2O$ (40 mL). The solid was dried in a vacuum oven at 40 °C to give 2.39 g (79%) of the amide.

Reference: Faul, M. M.; Winneroski, L. L.; Krumrich, C. A. *J. Org. Chem.* **1999**, *64*, 2465–2470.

# 3

# Oxidation

## 3.1 Alcohol to Ketone Oxidation State

### 3.1.1 Activated Manganese Dioxide Oxidation

Activated manganese dioxide ($MnO_2$) reliably oxidizes acetylenic, allylic, and benzylic alcohols to aldehydes and ketones. Saturated primary and secondary alcohols are also oxidized, albeit more slowly. The two main concerns are the activity of the manganese dioxide and the slow filtration of salts after the reaction. Activated $MnO_2$ is available commercially or may be prepared.

To a solution of 15.3 g (37.5 mmol) of the alcohol in 150 mL of hexanes was added 60 g of activated $MnO_2$. The reaction mixture was stirred at 22 °C overnight and filtered, and the solid residue was washed with 30% EtOAc in hexanes solution. The combined filtrates were dried ($Na_2SO_4$) and concentrated in vacuo. The residue was purified by chromatography on $SiO_2$ (EtOAc:hexanes, 1:10) to give 13.7 g (90%) of the ketone as a colorless oil.

Reference: Wipf, P.; Xu, W. *J. Org. Chem.* **1996**, *61*, 6556–6562.

### 3.1.2 Chromium-Based Oxidations

Chromium-based oxidations are reliable and well established, but the toxicity associated with chromium salts have meant that they are generally considered the

### 3.1.2.1 Pyridinium Chlorochromate

PCC: pyridinium with CrO$_3$Cl$^-$

Reaction: alcohol (cyclopentene with dioxolane and hydroxyalkyl chain) → aldehyde
Conditions: PCC, NH$_4$OAc, MS 4 Å, CH$_2$Cl$_2$, rt, 3 h, 63%

To a mixture of pyridinium chlorochromate (PCC 339 mg, 1.57 mmol), ammonium acetate (215 mg, 2.62 mmol), and 4 Å molecular sieves (610 mg) in CH$_2$Cl$_2$ (33 mL) was added a solution of the alcohol (208 mg, 1.05 mmol) in CH$_2$Cl$_2$ (14 mL) under argon at 0 °C over a period of 10 min. After the mixture had been stirred at room temperature for 3 h, diethyl ether (200 mL) was added and the mixture was filtered through a short pad of Florisil. The filtrate was washed successively with water (100 mL) and brine (100 mL), dried with Na$_2$SO$_4$, and concentrated. The residue was purified by chromatography on silica gel (hexane 70%, Et$_2$O 30%) followed by distillation to give the aldehyde as a colorless oil (132 mg, 63%).

Reference: Ohkita, M.; Kawai, H.; Tsuji, T. *J. Chem. Soc. Perkin. 1*, **2002**, 366–370.

### 3.1.2.2 Pyridinium Dichromate

PDC: (pyridinium)$_2$ with Cr$_2$O$_7$Cl$^{-2}$

Reaction: methyl oleandroside (alcohol) → ketone
Conditions: PDC, 3 Å sieves, AcOH, CH$_2$Cl$_2$, 99%

Methyl oleandroside (7.10 g, 40.4 mmol, 1.0 equivalents) in CH$_2$Cl$_2$ (200 mL) was treated with 3 Å powdered molecular sieves (20 g) and pyridinium dichromate (PDC 16.7 g, 44.3 mmol, 1.1 equivalents) at 2 °C followed by the addition of AcOH (4.0 mL). The mixture was warmed and stirred for 2 h at 25 °C. The mixture was treated with Celite (20 g), stirred for 30 min, and filtered. The solution was evaporated to a dark oil.

A solution of the oil in EtOAc (50 mL) was filtered through silica gel (50 g, 230–400 mesh) and the eluent was evaporated to give the ketone (7.0 g, 99%) as an oil.

Reference: Loewe, M. F.; Cvetovich, R. J.; DiMichele, L. M.; Shuman, R. F.; Edward J. J.; Grabowski, E. J. J. *J. Org. Chem.* **1994**, *59*, 7870–7875.

### 3.1.2.3 Collins Oxidation

The diamine (0.388 g, 1.0 mmol) and 4-methoxy-3-methoxycarbonylbenzoyl chloride (0.258 g, 2.1 mmol) were stirred in dry pyridine (5 mL) and $CH_2Cl_2$ (5 mL) overnight under nitrogen. Solid $CrO_3$ (0.6 g, 6.0 mmol) was added, and the mixture was stirred at 0 °C for 3 h. The $CH_2Cl_2$ was removed in vacuo and the solution was poured into water (100 mL). The solid produced was filtered and air-dried to give crude the ketone (1.2 g). Purification by silica gel column chromatography (20 g, 230–400 mesh), eluting with EtOAc:hexanes (2:1 v/v), gave the product as an amorphous glass (0.45 g, 85%).

Reference: Ruell, J. A.; De Clercq, E.; Pannecouque, C.; Witvrouw, M.; Stup, T. L.; Turpin, J. A.; Buckheit, Jr., R. W.; Cushman, M. *J. Org. Chem.* **1999**, *64*, 5858–5866.

### 3.1.3 2,3-Dichloro-5,6-dicyano-*p*-benzoquinone

Most aliphatic alcohols react slowly if at all with 2,3-dichloro-5,6-dicyano-*p*-benzoquinone (DDQ), allowing selective allylic or benzylic alcohol oxidation.

2,3-Dichloro-5,6-dicyano-*p*-benzoquinone (DDQ 908 mg, 4 mmol) was added to a solution of 4-hydroxybenzyl alcohol (496 mg, 4 mmol) in dioxane (24 mL). The reaction mixture immediately turned deep green (exothermic reaction), and $DDQH_2$ started precipitating within 1 min. Thin layer chromatography (TLC) analysis indicated consumption of starting material after 15 min. The solvent was removed from the yellow reaction mixture in vacuo. Treatment of the residue with $CH_2Cl_2$ left $DDQH_2$ undissolved (quantitatively). Filtration followed by evaporation of $CH_2Cl_2$ gave 4-hydroxybenzaldehyde (74% yield) which was recrystallized from water.

Reference: Becker, H.-D.; Bjork, A.; Alder, E. *J. Org. Chem.* **1980**, *45*, 1596–1600.

### 3.1.4 Dimethylsulfoxide-Based Oxidations

"Those who've learned use the Swern" (John L. Wood). The Swern oxidation is a very reliable oxidation, and is likely the oxidation of choice when the temperature can be carefully controlled (see Tidwell, T. T. *Org. React.* **1990**, *39*, 297–572). However, the active chlorosulfonium intermediate decomposes above approximately −60 °C. Therefore, alternative dehydrating agents have been used (for useful reviews see Mancuso, A. J.; Swern, D. *Synthesis* **1981**, 165; Lee, T. V. Oxidation Adjacent to Oxygen of Alcohols by Activated DMSO Methods. In *Comprehensive Organic Synthesis*; Trost, B. M., Ed.; Pergamon Press: Oxford, 1991; Vol. 7, 291–304). A representative list is shown below.

#### 3.1.4.1 With Oxalyl Chloride (Swern)

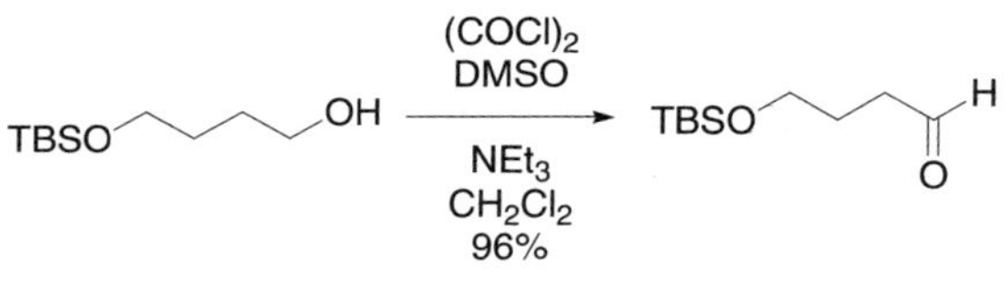

$$TBSO\sim\sim\sim OH \xrightarrow[\substack{NEt_3 \\ CH_2Cl_2 \\ 96\%}]{\substack{(COCl)_2 \\ DMSO}} TBSO\sim\sim\sim CHO$$

To a solution of oxalyl chloride (2.1 mL, 24 mmol, 1.2 equivalents) in $CH_2Cl_2$ (30 mL) cooled at −78 °C was added dropwise a solution of dimethylsulfoxide (DMSO 3.3 mL, 21 mmol, 1.1 equivalents) in $CH_2Cl_2$ (32 mL). After 5 min, a solution of 4-(*tert*-butyldimethylsilanyloxy)butan-1-ol (4.0 g, 20 mmol, 1.0 equivalents) in $CH_2Cl_2$ (26 mL) was added. The reaction mixture was then stirred for 15 min at −78 °C and triethylamine (14.0 mL, 100 mmol, 5.00 equivalents) was added in 1 portion. After 10 min at −78 °C, the mixture was allowed to warm to room temperature and diluted with $CH_2Cl_2$ (140 mL). The organic layer was successively washed with a saturated aqueous solution of $NH_4Cl$ (30 mL) and brine (2 × 30 mL). The combined organic extracts were dried over $MgSO_4$, filtered, and concentrated under reduced pressure. Purification by flash chromatography on silica gel (petroleum ether/EtOAc: 90/10) afforded 3.88 g (96% yield) of the aldehyde as a colorless oil.

Reference: Taillier, C.; Gille, B.; Bellosta, V.; Cossy, J. *J. Org. Chem.* **2005**, *70*, 2097–2108.

### 3.1.4.2 With Dicyclohexylcarbodiimide (Moffatt)

To a solution of the alcohol (31.0 g, 0.10 mol), dicyclohexylcarbodiimide (DCC 42.0 g, 0.21 mol), DMSO (18.1 mL, 0.25 mol), and pyridine (17.2 mL, 0.10 mol) in toluene (500 mL) was added trifluoroacetic acid (8.3 mL, 0.10 mol) dropwise at 0 °C over 10 min. After being stirred at room temperature for 10 h, the resulting suspension was filtered through a Celite pad. The filtrate was washed with $H_2O$, saturated $NaHCO_3$, and brine, dried over $MgSO_4$, filtered, and concentrated in vacuo. The residue was purified by column chromatography on a silica gel (EtOAc:hexane, 1:30) to give a crude compound contaminated by reduced DCC which was dissolved in hexane. The precipitate was filtered off, and the filtrate was concentrated in vacuo. The residue was re-purified by silica gel column chromatography (EtOAc:hexane = 1:30), giving the ketone (23.4 g, 75%).

Reference: Jin, Y. H.; Liu, P.; Wang, J.; Baker, R.; Huggins, J.; Chu, C. K. *J. Org. Chem.* **2003**, *68*, 9012–9018.

### 3.1.4.3 With $SO_3$·Pyridine (Parikh–Doering Oxidation)

A 640-L vessel was charged with $SO_3$·pyridine (51.47 kg, 323.1 mol). DMSO (169 L) was added and the whole was heated slowly to 33 °C. After a solution was obtained, it was cooled to 25 °C. The TsOH salt of the amine (40.9 kg, 107.7 mol) was added into the vessel and suspended in DMSO (50 L). After the addition of $Et_3N$ (62 L, 43.8 mol), the $SO_3$·pyridine solution in DMSO was added to the two-phase mixture in the vessel at such a rate as to keep the internal temperature below 25 °C. After 1 h of stirring at 22 °C, the reaction was 92% complete. The mixture was cooled to 10 °C and quenched with water (182 L) over a period of 40 min at such a rate as to keep the internal temperature below 17 °C; a 25% $NH_3$ solution (16 L) was then added. After phase separation, the aqueous phase was extracted with three portions of toluene (3 × 60 L) while controlling the pH of the aqueous layer to 10 after each extraction. The combined organic phases (approximately 240 L) were extracted with water (61 L), and the bright-orange solution was heated at 40–50 °C jacket temperature over 1 h while blowing a nitrogen stream into the solution via an immersing tube. Then, toluene (170 L) was stripped off at 50 °C to afford the ketone (50.84 kg, 93%) as an orange solution in toluene.

Reference: Ripin, D. H. B.; Abele, S.; Cai, W.; Blumenkopf, T.; Casavant, J. M.; Doty, J. L.; Flanagan, M.; Koecher, C.; Laue, K. W.; McCarthy, K.; Meltz, C.; Munchhoff, M.; Pouwer, K.; Shah, B.; Sun, J.; Teixeira, J.; Vries, T.; Whipple, D. A.; Wilcox, G. *Org. Process Res. Dev.* **2003**, *7*, 115–120.

### 3.1.4.4 With Trifluoracetic Anhydride

The crude alcohol (5.283 g, 94 area% pure by high performance liquid chromatography, 10.22 mmol) was suspended in dry tetrahydrofuran (24.39 g). Dry DMSO was added dropwise until a homogeneous solution was achieved (4.261 g, 54.6 mmol), and the resulting solution was cooled to −15 °C. Trifluoroacetic anhydride (3.450 g, 16.43 mmol) was added dropwise over 0.25 h. No significant heat liberation was noted. The colorless, homogeneous solution was stirred at −15 °C for 0.5 h after which dry triethylamine (4.44 g, 43.9 mmol) was added over 0.1 h. During this addition, the reaction temperature increased from −15 °C to −10 °C. The pale yellow reaction mixture was warmed to room temperature and stirred for 1 h. The mixture was then slowly added to cold water (100 mL) with vigorous stirring. The resulting precipitate was collected by suction filtration, and the filter cake was thoroughly washed with water (50 mL). The filter cake was dried in vacuo leaving a tan, free-flowing powder (4.35 g, 88%).

Reference: Appell, R. B.; Duguid, R. J. *Org. Process Res. Dev.* **2000**, *4*, 172–174.

### 3.1.4.5 With Acetic Anhydride

To a mixture of alcohol (886 g, 1.0 equivalents, 2.5 mol) and DMSO (7.55 L) was added acetic anhydride (5.05 L, 21.4 equivalents, 53.4 mol). The mixture was stirred at room temperature for 18 h. The mixture was diluted with ethanol (16.8 L), stirred for 1 h, and diluted with water (4.2 L). Ammonium hydroxide (11 L) was added while maintaining the temperature at 15–30 °C by cooling and the mixture was then diluted with water (16.8 L). Filtration gave a solid which was washed with water and dried to give the ketone (818 g, 93%) as a tan solid.

Reference: Albright, J. D.; Goldman, L. *J. Org. Chem.* **1967**, *89*, 2416–2423.

### 3.1.4.6 With 2,4,6-Trichloro-[1,3,5]-triazine

Dimethylsulfoxide (1.25 mL, 17.6 mmol) was added to a solution of 2,4,6-trichloro-[1,3,5]-triazine (TCT, 0.66 g, 3.6 mmol) in THF (20 mL), stirred, and maintained at −30 °C. After 30 min, *N*-benzyloxycarbonyl-2-amino-3-phenylpropan-1-ol, (0.86 g, 3 mmol) in THF (10 mL) was added slowly at −30 °C with stirring, and after an additional 30 min, NEt$_3$ (2 mL, 14.3 mmol) was added. After 15 min, the mixture was warmed to room temperature, the solvent evaporated in vacuo, and Et$_2$O (50 mL) added to the resulting solid. The mixture was quenched with 1 N HCl, and the organic phase washed with 15 mL of a saturated solution of NaHCO$_3$, followed by brine. The organic layer was dried (Na$_2$SO$_4$) and the solvent evaporated to yield pure *N*-benzyloxycarbonyl-2-amino-3-phenylpropionaldehyde (0.77 g, 90%).

Reference: De Luca, L.; Giacomelli, G.; Porcheddu, A. *J. Org. Chem.* **2001**, *66*, 7907–7909.

### 3.1.4.7 With P$_2$O$_5$

Anhydrous DMSO (650 mL) was cooled to 18–20 °C under nitrogen in a 3-L round-bottomed glass flask. DMSO solidifies at 18 °C and therefore it is important to keep the reaction mixture just above freezing point. To this cold solution was added P$_2$O$_5$ (142 g, 1.0 mol, 1 equivalent) in 3 portions under a N$_2$ atmosphere. The addition of P$_2$O$_5$

to DMSO is exothermic, and if the mass temperature exceeds 28 °C, the color darkens and the product will be of inferior quality. The mixture was cooled to 18–20 °C between each addition. After addition of $P_2O_5$ was completed, the mixture was stirred at 18–25 °C for 10–15 min. 1,2:5,6-Di-$O$-isopropylidene-$D$-glucofuranose (260 g, 1.0 mol) was dissolved in anhydrous DMSO (1.3 L) and added over 30 min (maintaining the temperature at 18–25 °C) to the stirred solution of $P_2O_5$ in DMSO under a $N_2$ atmosphere. The resulting solution was heated to 50–55 °C for 3 h. TLC (eluent: $CH_2Cl_2$:MeOH, 95:5) shows complete conversion of glucofuranose ($R_f$ = 0.68) to ulose ($R_f$ = 0.81). The reaction mixture was allowed to reach 25–30 °C and was extracted twice with methyl *tert*-butyl ether (MTBE 1.5 and 1 L) in a 6-L separation funnel. The combined MTBE layer (~4 L) was concentrated in vacuo (water-bath temperature set to 40 °C) to approximately 2 L and allowed to reach 25–30 °C. $NaBH_4$ (24 g, 0.63 mol) was dissolved in water (1 L, 55.6 mol) at 0–10 °C, and the concentrated MTBE layer was added to the aqueous layer over 30 min to keep the temperature at 0–10 °C. TLC (eluent: EtOAc/heptane, 6:4) after 30 min shows full conversion of ulose ($R_f$ = 0.53) to 1,2:5,6-di-$O$-isopropylidene-$D$-allofuranose ($R_f$ = 0.39). The reaction mixture was allowed to reach 25–30 °C. $CH_2Cl_2$ (1 L) and water (500 mL) were added, and the layers were separated. The aqueous layer was extracted once more with $CH_2Cl_2$ (500 mL). The combined organic layers were concentrated in vacuo to an oil which was subsequently dissolved in MTBE (300 mL) and extracted with water (3 × 500 mL). The combined aqueous layers were extracted with $CH_2Cl_2$ (3 × 500 mL). The combined $CH_2Cl_2$ layers were dried ($Na_2SO_4$, 100 g), filtered, and concentrated in vacuo to provide the crude oil. Crystallization from cyclohexane (500 mL), washing of crystals with cold *n*-pentane, and drying hereof in vacuo afforded analytically pure 1,2:5,6-di-$O$-isopropylidene-$D$-allofuranose (191 g, 73%).

Reference: Christensen, S. M.; Hansen, H. F.; Koch, T. *Org. Proc. Res. Dev.* **2004**, *8*, 777–780.

### 3.1.4.8 With N-Chlorosuccinimide (Corey–Kim)

The Corey–Kim reaction is different from DMSO-based oxidations in that the first step is an oxidation of dimethyl sulfide, rather than activation of DMSO. The active oxidizing agent, however, is the chlorosulfonium ion analogous to the Swern oxidation.

The alcohol (1.5 L, 208 g, 280 mmol) in THF is charged to a jacketed flask followed by dimethyl sulfide (37 g, 590 mmol) and diisopropylethylamine (47 g, 364 mmol). The solution is cooled to approximately −13 °C. N-Chlorosuccinimide (NCC 71 g, 532 mmol) is dissolved in THF (240 mL) and added to the flask at a rate so as to maintain the internal temperature at −11 to −13 °C. The mixture is then stirred at −15 ± 5 °C for 3 h. Isopropyl acetate (3 L) is added followed by 0.5 N NaOH (1.2 L). The mixture is warmed to room temperature and stirred for 1 h. The organic layer is washed with 5% NaCl (2 × 600 mL) and brine (2 × 600 mL). The product layer is concentrated under vacuum to obtain, at first, a yellow amorphous solid which when dried under high vacuum turns into a white foam. Slurrying the solid in warm water followed by filtration and drying afforded 196 g (94%) of the 3-ketomacrolide as a white solid after trituration with 10% EtOAc/heptane.

Reference: Kerdesky, F. A. J.; Premchandran, R.; Wayne, G. S.; Chang, S.-J.; Pease, J. P.; Bhagavatula, L.; Lallaman, J. E.; Arnold, W. H.; Morton, H. E.; King, S. A. *Org. Proc. Res. Dev.* **2002**, *6*, 869–875.

### 3.1.5 Fleming Oxidation

To a suspension of the siloxane (760.3 mg, 1.0 mmol) and KHCO$_3$ (300.2 mg, 3.0 mmol) in a 1:1 mixture of MeOH:THF (10 mL) was added H$_2$O$_2$ (0.76 mL, 30% aqueous, 5 mmol). After being heated to 53 °C, the mixture was cooled to 0 °C (ice bath), and Na$_2$S$_2$O$_3$ (2.80 g, 17.7 mmol, 17.7 equiv) was added in 4 portions over 40 minutes. The mixture was then allowed to warm to room temperature for 2 h until H$_2$O$_2$ could not be detected by iodine/starch test paper. The inorganic salts were removed by filtration (Celite) and were washed with ether (25 mL). Removal of the solvent in vacuo provided a colorless oil which was purified twice by column chromatography (SiO$_2$, hexane/EtOAc 4/1 → 1/1; then hexane/EtOAc 1/2) to provide the diol (595.8 mg, 90%) as a colorless viscous oil.

Reference: Denmark, S. E.; Cottell, J. J. *J. Org. Chem.* **2001**, *66*, 4276–4284.

### 3.1.6 Iodosobenzene-Based Oxidations

#### 3.1.6.1 Dess–Martin periodinane

The Dess–Martin (D–M) oxidation is the method of choice for the oxidation of alcohols bearing sensitive functional groups to the corresponding carbonyl compounds.

The reagent that accomplishes this oxidation is 1,1,1-triacetoxy-1,1-dihydro-1,2-benziodoxol-3(1*H*)-one or simply the Dess–Martin periodinane (DMP). Several reviews have been written (see Moriarty, R. M.; Prakash, O. *Org. React.* **1999**, *54*, 273–418; De Munari, S.; Frigerio, M.; Santagostino, M. *J. Org. Chem.* **1996**, *61*, 9272–9279). The preferred preparation of DMP is found in Meyer, S. D.; Schreiber, S. L. *J. Org. Chem.* **1994**, *59*, 7549–7552.

### 3.1.6.1.1 Standard

Dess–Martin periodinane (579 mg, 1.37 mmol, for preparation see chapter 1) was added to a solution of oxazole alkynol (214 mg, 0.44 mmol) in 25 mL of $CH_2Cl_2$. After 30 min, saturated aqueous $NaHCO_3$ and excess $Na_2S_2O_3$ were added to the reaction mixture. After the solids were dissolved, the mixture was extracted with $CH_2Cl_2$. The combined organic layers were washed with saturated aqueous $NaHCO_3$, dried ($MgSO_4$), and filtered. After removal of solvent (aspirator), the residue was purified by flash chromatography on silica gel (13 mm × 20 cm) using 2:3 EtOAc/hexane as the eluent, to afford 102 mg (48%) of the ketone as an oil.

Reference: Vedejs, E.; Piotrowski, D. W.; Tucci, F. C. *J. Org. Chem.* **2000**, *65*, 5498–5505.*

*For substrates with simpler functionalities, the yields are generally in the range of > 90%.

### 3.1.6.1.2 Schreiber–Meyer Modification

The Dess–Martin oxidation reaction rate is significantly accelerated in the presence of 1.1 equivalents of water (see table below). In addition, excess reagent and water do not appear to increase the reaction rate. In the presence of a controlled amount of water, the Dess–Martin reagent is partially hydrolyzed to an acetoxyiodinane oxide, which is a more reactive oxidant than DMP, thus an increase in the reaction rate is seen. To obtain reproducible results, the DMP should be of high purity prior to the addition of water.

acetoxyiodinane oxide

OXIDATION 65

| DMP (eq) | H₂O (eq) | reaction time (h) | Yield |
|---|---|---|---|
| 1.5 | 0 | 14 | 97% |
| 1.5 | 1.1 | 0.5 | 97% |
| 4.9 | excess | 1.2 | 98% |

Water (10 mL, 0.55 mmol) was solvated in $CH_2Cl_2$ (10 mL) by drawing the solvent mixture into and expelling it from a disposable pipette several times. The wet $CH_2Cl_2$ was added slowly via a dropping funnel to a vigorously stirring solution of *trans*-2-phenylcyclohexanol (88.4 mg, 0.502 mmol) and DMP (321 mg, 0.502 mmol) in dry $CH_2Cl_2$ (3 mL). The clear solution grew cloudy toward the end of the wet $CH_2Cl_2$ addition, which required 30 min. The mixture was diluted with ether and then concentrated into a few milliliters of solvent by rotary evaporator. The residue was taken up in ether and then washed with a mixture of 10% $Na_2S_2O_3$ and saturated aqueous $NaHCO_3$ (1:1, 15 mL), followed by water (10 mL) and brine (10 mL). The aqueous washes were back-extracted with ether (20 mL), and this organic layer was washed with water and brine. The combined organic layers were dried ($Na_2SO_4$), filtered, and concentrated. The crude product was purified by flash chromatography eluting with hexane:EtOAc (20:1 to 10:1) to provide 84.7 mg (97%) of 2-phenylcyclohexanone as a crystalline solid.

Reference: Meyer, S. D.; Schreiber, S. L. *J. Org. Chem.* **1994**, *59*, 7549–7552.

### 3.1.6.2. 1-Hydroxy-1,2,benziodoxol-3(1H)-one Oxidation

Piperonyl alcohol (0.15 g, 1.00 mmol) was dissolved in EtOAc (7 mL, 0.14 M final concentration), and 1-hydroxy-1,2,benziodoxol-3(1*H*)-one (IBX 0.84 g, 3.00 mmol) was added. The resulting suspension was immersed in an oil bath set to 80 °C and stirred vigorously open to the atmosphere. After 3.25 h (TLC monitoring), the reaction was cooled to room temperature and filtered through a medium glass frit. The filter cake was washed with 3 × 2 mL of EtOAc, and the combined filtrates were concentrated to yield 0.14 g (90%, > 95% pure by ¹H NMR) of piperonal as a waxy solid.

Reference: More, J. D.; Finney, N. S. *Org. Lett.* **2002**, *4*, 3001–3003.

## 3.1.7 Oppenauer Oxidation

A solution of the alcohol (40.1g, 0.23 mol) in toluene (180 mL) and acetone (105 mL) containing aluminum isopropoxide (7.7 g, 0.038 mol) was heated to reflux (87 °C) under argon for 4.5 h. The cooled solution was then treated with water (77 mL) and filtered. The aqueous layer was extracted with ether ($3 \times 250$ mL) and the organic layers were combined, washed with water (250 mL), dried ($Na_2SO_4$), filtered, and concentrated in vacuo. The resulting oil can be distilled under reduced pressure, but can also be crystallized directly in the refrigerator under argon after addition of petroleum ether (40–60, 50 mL) to give the ketone (34.0 g, 87%) as a white solid.

Reference: Bagal, S. K.; Adlington, R. M.; Baldwin, J. E.; Marquez, R.; Cowley, A. *Org. Lett.* **2003**, *5*, 3049–3052.

### 3.1.8 Pummerer Rearrangement

The Pummerer rearrangement is similar mechanistically to a DMSO-based oxidation, transferring oxidation from a sulfur to a carbon, in this case intramolecularly. Reviews of the Pummerer rearrangement and its variants have been written: (a) Bur, S. K.; Padwa, A. *Chem. Rev.* **2004**, *104*, 2401–2432. (b) Padwa, A.; Waterson, A. G. *Curr. Org. Chem.* **2000**, *4*, 175–203. (c) Padwa, A.; Gunn, Jr., D. E.; Osterhout, M. H. *Synthesis* **1997**, 1353–1377. (d) DeLucchi, O.; Miotti, U.; Modena, G. *Org. React.* **1991**, *40*, 157–405.

A 10-mL round-bottomed flask equipped with a rubber septum and argon inlet needle was charged with a solution of the sulfoxide (91 mg, 0.35 mmol) in 3 mL of toluene. Acetic anhydride (0.164 mL, 0.177 g, 1.74 mmol) and *p*-toluenesulfonic acid (2 mg, 0.001 mmol) were added, and the flask was fitted with a reflux condenser and heated at reflux for 1 h. The resulting mixture was cooled to 25 °C and concentrated to afford 113 mg of an oil. Column chromatography on silica gel (gradient elution with 5-10% ethyl acetate-hexane) provided the mixed acetal (94 mg, 89%) as a colorless oil.

Reference: Lawlor, M. D.; Lee, T. W.; Danheiser, R. L. *J. Org. Chem.* **2000**, *65*, 4375–4384.

### 3.1.9 2,2,6,6-Tetramethylpiperdine 1-oxyl-Catalyzed Oxidation

There are a variety of stoichiometric oxidants that can be used in conjunction with 2,2,6,6-tetramethylpiperdine 1-oxyl (TEMPO), including TCCA, iodobenzene diacetate, *meta*-chloroperbenzoic acid (*m*-CPBA), sodium bromite, sodium hypochlorite, and *N*-chlorosuccinimide. TEMPO can be used to oxidize primary alcohols to aldehydes or carboxylic acids, based on reaction conditions. A particularly attractive feature is that the product is usually isolated quite pure after concentration of the organic layer.

The following is typical—the aldehyde is used crude, and therefore the yield for the TEMPO oxidation is combined with the yield in the next step.

TEMPO

TEMPO, NaOCl, NaHCO$_3$ (aq), CH$_2$Cl$_2$, 84%

To the diol (735 mg, 3.93 mmol) in CH$_2$Cl$_2$ (35 mL) at 0 °C was added TEMPO (13 mg, 0.080 mmol) and KBr (47 mg, 0.39 mmol). The mixture was vigorously stirred and NaOCl (approximately 1.5 M in H$_2$O, 4.0 mL, 5.9 mmol) in 25-mL pH 8.6 buffer (0.5 M NaHCO$_3$/0.05 M Na$_2$CO$_3$) was added in portions. Additional NaOCl solution was added in portions until the reaction maintained a dark color and TLC analysis indicated the complete consumption of the starting material. The reaction was quenched by addition of MeOH (1 mL) and the aqueous phase extracted with CH$_2$Cl$_2$ (3 × 30 mL). The combined organic solutions were washed with brine (60 mL), dried over Na$_2$SO$_4$, and concentrated under reduced pressure to provide the aldehyde which was used without further purification. The yield for the aldehyde formation and next step (oxime formation) was 84%.

Reference: Bode, J. W.; Carreira, E. M. *J. Org. Chem.* **2001**, *66*, 6410–6424.

### 3.1.10 Tetrapropylammonium Perruthenate oxidation

Discovered in 1987, the tetra-*n*-propylammonium perruthenate (TPAP) oxidation is straightforward to run. The reagent is expensive and thus it is used in catalytic amounts and becomes re-oxidized by *N*-methyl morpholine-*N*-oxide. For a review see Ley, S. V.; Norman, J.; Griffith, W. P.; Marsden, S. P. *Synthesis* **1994**, 639–666.

TPAP, NMO, 4 Å MS, CH$_2$Cl$_2$, 90%

Tetrapropylammonium perruthenate (TPAP 63 mg) was added in 1 portion to a stirred mixture of $N^\alpha$-Boc-4-*trans*-hydroxy-L-proline (1 g, 3.47 mmol), *N*-methyl-morpholine *N*-oxide (0.62 g, 15.6 mmol), and powdered molecular sieves (4 Å, 1.78 g) in CH$_2$Cl$_2$ (7 mL) at room temperature under argon. The mixture was stirred for 3 h, filtered, and evaporated in vacuo to give a black residue. The product was purified by flash column chromatography on silica gel (CH$_2$Cl$_2$:EtOAc = 1:1, v/v) and recrystallized from ether and hexane to give the corresponding ketone (0.9 g, 90%).

Reference: Tamaki, M.; Han, G.; Hruby, V. J. *J. Org. Chem.* **2001**, *66*, 3593–3599.

### 3.1.11 Wacker Oxidation

The Wacker oxidation oxidizes a terminal olefin to a methyl ketone.

Review: Takars, J. M.; Jiang, X.-t. *Curr. Org. Chem.* **2003**, *7*, 369–396.

A suspension of PdCl$_2$ (4 mg, 23.6 μmol) and CuCl (23 mg, 0.23 mmol) in a mixture of DMF and H$_2$O (7:1, 0.3 mL) was stirred under oxygen atmosphere at room temperature for 1 h. A solution of the olefin (79 mg, 0.24 mmol) in DMF and H$_2$O (7:1, 0.1 mL) was added to the reaction mixture. After being stirred at room temperature overnight, the mixture was quenched with 20% KHSO$_4$ and extracted with Et$_2$O three times. The extracts were successively washed with saturated NaHCO$_3$ and brine, dried, and evaporated. The residue was purified with chromatography using *n*-hexane:EtOAc (3:1) as eluent to yield the ketone (71 mg, 86%).

Reference: Takahata, H.; Ouchi, H.; Ichinose, M.; Nemoto, H. *Org. Lett.* **2002**, *4*, 3459–3462.

## 3.2 Alcohol to Acid Oxidation State

For a review of chromium–amine complex oxidations, see Luzzio, F. A. *Org. React.* **1998**, *53*, 1–221.

### 3.2.1 Jones Oxidation

To a stirred solution of the alcohol in acetone (~ 0.1 M) at 0 °C was added Jones reagent (8 N, see chapter 1 for the preparation of this reagent) until the solution remained orange. The reaction was left to stir for 10 min before water (40 mL) was added, and the aqueous layer was extracted with hexanes (4 × 30 mL). The combined organic layers were washed with brine (50 mL), dried over MgSO$_4$, and concentrated in vacuo to afford the acid at 89% yield.

## 3.2.2 Permanganate Oxidation

$$\text{F}_3\text{C-CF}_2\text{-CH}_2\text{CH}_2\text{CH}_2\text{OH} \xrightarrow[\text{H}_2\text{O}]{\text{NaMnO}_4, \text{Et}_4\text{NHSO}_4} \text{F}_3\text{C-CF}_2\text{-CH}_2\text{CH}_2\text{COOH}$$
77%

4,4,5,5,5-Pentafluoropentanol (1.8 kg, 10.1 mol), tetraethylammonium hydrogen sulfate (18.1 g, 0.08 mole), and water (10.8 L) were added to a 50 L QVF vessel and heated with stirring to 70 °C. Sodium permanganate monohydrate (2.33 kg, 14.14 mol) was dissolved at 20 °C in water (10.8 L) and transferred to a measure vessel. Aqueous sodium permanganate was added in aliquots (approximately 10% at a time) to the stirred aqueous solution of pentafluoropentanol and tetraethylammonium hydrogen sulfate maintaining a temperature of 65–75 °C by the additions of permanganate. The total time taken to add the aqueous permanganate was 2 h 30 min. The reaction was stirred at 70 °C for 4 h when gas chromatography (GC) analysis showed conversion of pentafluoropentanol to pentafluoropentanoic acid to be complete. The reaction mixture was allowed to cool to ambient temperature overnight and screened through a Celite filter aid (500 g) to remove precipitated manganese dioxide. The isolated manganese dioxide was washed with hot water (60 °C, 18 L). The combined aqueous layers were extracted with methyl *tert*-butyl ether (5.4 L), and the upper organic layer was discarded. The aqueous layer was acidified with concentrated sulfuric acid (320 mL) to pH 1. The *lower* organic layer which separated was retained. The aqueous layer was extracted with methyl *tert*-butyl ether (2 × 5.4 L), and the upper organic layers were combined with the initial, lower organic layer. The combined organic layers were washed with water (5.4 L) and dried with anhydrous sodium sulfate. The organic solvent was removed in vacuo at 50 °C and the residue distilled to give the acid as a pale pink, low-melting solid (1.49 kg, 77%).

Reference: Mahmood, A.; Robinson, G. E.; Powell, L. *Org. Proc. Res. Dev.* **1999**, *3*, 363–364.

## 3.2.3 2,2,6,6-Tetramethylpiperdine 1-oxyl- Catalyzed Oxidation

A mixture of the alcohol (11.4g, 40 mmol), TEMPO (436 mg, 2.8 mmol), MeCN (200 mL), and sodium phosphate buffer (150 mL, 0.67 M, pH = 6.7) is heated to 35 °C. Sodium chlorite (NaClO$_2$, 9.14 g 80%, 80.0 mmol in 40 mL water) and dilute bleach (1.06 mL 5.25% NaOCl diluted into 20 mL, 2.0 mol %) are then added simultaneously over 2 h. (Caution! Do not mix bleach and $NaClO_2$ before being added to the reaction mixture.) The mixture is stirred at 35 °C until the reaction is complete and then cooled to room temperature. Water (300 mL) is added, and the pH is adjusted to 8.0 with 2.0 N NaOH (48 mL). The reaction is quenched by pouring into cold (0 °C) $Na_2SO_3$ solution (12.2 g in 200 mL water) maintained at < 20 °C. The pH of the aqueous layer should be 8.5–9.0. After stirring for 0.5 h at room temperature, MTBE (200 mL) is added. The organic layer is separated and discarded. More MTBE (300 mL) is added, and the aqueous layer is acidified with 2.0 N HCl (100 mL) to pH 3–4. The organic layer is separated, washed with water (2 × 100 mL) and brine (150 mL), and then concentrated to give the crude Cbz-phenylalanine (10.2 g, 85%) with no detectable racemization.

Reference: Zhao, M.; Li, J.; Mano, E.; Song, Z.; Tschaen, D. M.; Grabowski, E. J. J.; Reider, P. J. *J. Org. Chem.* **1999**, *64*, 2564–2566.

## 3.3 Olefin to Diol

### 3.3.1 Dimethyl Dioxirane

*Caution*: Dioxiranes are usually volatile peroxides and thus should be handled with care by observing all safety measures. The preparations and oxidations should be carried out in a hood with good ventilation. *Inhalation and direct exposure to skin must be avoided!* There is an excellent review of dioxirane (DMDO) chemistry in Adam, W.; Saha-Möller, C. R.; Zhao, C.-G. *Org. React.* **2002**, *61*, 219–516.

*Preparation of Dimethyl Dioxirane*

$$\underset{\text{H}_2\text{O}}{\overset{\text{NaHCO}_3, \text{ oxone}}{\xrightarrow{\hspace{2cm}}}}$$

A 2-L, three-necked, round-bottomed flask containing a mixture of water (80 mL), acetone (50 mL, 0.68 mol), and sodium bicarbonate (96 g) is equipped with a magnetic stir bar and a pressure equalizing addition funnel containing water (60 mL) and acetone (60 mL, 0.82 mol). A solid addition flask containing Oxone (180 g, 0.29 mol) is attached to the reaction vessel via a rubber tube. An air condenser (20 cm length) loosely packed with glass wool is attached to the reaction vessel. The outlet of the air condenser is connected to a 75 × 350-mm Dewar condenser filled with dry ice–acetone that is connected to a receiving flask (100 mL) cooled in a dry ice–acetone bath. The receiving flask is also connected in series to a second dry ice–acetone cold trap, a trap containing a potassium iodide solution, and a drying tube. A gas inlet tube is connected to the reaction flask and a stream of nitrogen gas is bubbled through the reaction mixture. The Oxone is added in portions (10–15 g) while the acetone–water mixture is simultaneously added dropwise. The reaction mixture is stirred vigorously throughout the addition of reagents (approximately 30 min). A yellow solution of dimethyldioxirane in acetone collects in the receiving flask. Vigorous stirring is continued for an additional 15 min while a slight vacuum (about 30 mm, water aspirator) is applied to the cold trap. The yellow dioxirane solution (62–76 mL) is dried over sodium sulfate ($Na_2SO_4$), filtered, and stored in the freezer (−25°C) over $Na_2SO_4$. The dioxirane content of the solution is assayed using phenyl methyl sulfide and the gas-liquid chromatography (GLC) method. Generally, concentrations in the range of 0.07–0.09 M are obtained.

Reference: Murray, R. W.; Singh M. *Org. Syn.* **1998**, *Coll. Vol. IX*, 288–293.

*Epoxidation with Dimethyl Dioxirane*

To a magnetically stirred solution of *trans*-stilbene (0.724 g, 4.02 mmol) in 5 mL of acetone in a 125-mL stoppered Erlenmeyer flask was added a solution of DMDO in acetone (0.062 M, 66 mL, 4.09 mmol) at room temperature (about 20 °C). The progress of the reaction was followed by GLC analysis, which indicated that *trans*-stilbene was converted into the oxide in 6 h. Removal of the excess acetone on a rotary evaporator (20 °C, 15 mm Hg) afforded a white crystalline solid. The solid was dissolved in $CH_2Cl_2$ (30 mL) and dried over anhydrous $Na_2SO_4$. The drying agent was removed by filtration and washed with $CH_2Cl_2$. The solution was concentrated on a rotary evaporator, and the remaining solvent was removed (20 °C, 15 mmHg) to give an analytically pure sample of the oxide (0.788 g, 100%).

Reference: Murray, R. W.; Singh M. *Org. Syn.* **1998**, *Collective Volume 9*, 288–293.

### 3.3.2 Halogenation with *N*-bromosuccinimide

To a solution of the olefin (500 mg, 2.29 mmol, 1.0 equivalents) in THF and water (50 mL, 1:1 v:v) was added *N*-bromosuccinimide (NBS 2 g, 28.1 mmol, 12.3 equivalents) at room temperature and the mixture was stirred for 22 h. The reaction mixture was diluted with water (50 mL) and extracted with $Et_2O$ (2 × 100 mL), and the organic layer was washed with water (3 × 50 mL), dried ($MgSO_4$), and concentrated. Crystallization of the residue from ether/petroleum ether gave the bromohydrin as colorless crystals (550 mg, 76%).

Reference: Kutney, J. P.; Singh, A. K. *Can. J. Chem.* **1982**, *60*, 1842–1846.

### 3.3.3 Jacobsen–Katsuki Epoxidation

Jacobsen–Katsuki epoxidation works best with *cis*-styrene type olefins.

A solution of 0.05 M $Na_2HPO_4$ (10.0 mL) was added to a 25-mL solution of undiluted commercial household bleach (Clorox). The pH of the resulting buffered solution (0.55 M in NaOCl) was adjusted to pH 11.3 by addition of a few drops of 1 M NaOH solution. This solution was cooled to 0 °C and then added at once to a 0 °C solution of the manganese catalyst (260 mg, 0.4 mmol) and *cis*-o-methylstyrene (1.18 g, 10 mmol) in 10 mL of $CH_2Cl_2$. The two-phase mixture was stirred at room temperature, and the reaction progress was monitored by TLC. After 3 h, 100 mL of hexane was added to the mixture and the brown organic phase was separated, washed twice with 100 mL of $H_2O$ and once with 100 mL of saturated NaCl solution, and then dried ($Na_2SO_4$). After solvent removal, the residue was purified by flash chromatography on silica gel to afford the epoxide (0.912 g, 68%). The *ee* of the epoxide was determined to be 84% by $^1H$ NMR analysis in the presence of $Eu(hfc)_3$.

Reference: Zhang, W.; Jacobsen, E. N. *J. Org. Chem.* **1991**, *56*, 2296–2298.

### 3.3.4 Oxidation of Olefin with *meta*-Chloroperbenzoic Acid

*meta*-Chloroperbenzoic acid (*m*-CPBA) is known to decompose at elevated temperatures (starting at approximately 97 °C) and therefore reactions with *m*-CPBA should be run below 50 °C (see Kubota, A; Takeuchi, H. *Org. Proc. Res. Dev.* **2004**, *8*, 1076–1078).

A solution of *m*-CPBA (10.4 g, 70%, 42.3 mmol) in $CH_2Cl_2$ (80 mL) was added to a solution of the olefin (3.73 g, 19.2 mmol) in $CH_2Cl_2$ (40 mL). The reaction was stirred for 12 h and cooled to 0 °C. 2-Methyl-2-butene (8.1 mL, 76.9 mmol) was added (to quench the reaction) and the resulting mixture was slowly warmed to 25 °C and stirred for 4 h to consume excess m-CPBA. The mixture was diluted with saturated $NaHCO_3$ and extracted by $CH_2Cl_2$. The combined $CH_2Cl_2$ extracts were washed with saturated $Na_2SO_3$ (30 mL), 5% NaOH (2 × 30 mL), and water (2 × 30 mL), dried ($MgSO_4$), and concentrated to afford the crude epoxide. Flash chromatography on silica gel (10:1 hexanes/EtOAc) yielded 4.03 g (99%) of pure epoxide as a colorless oil.

Reference: Snider, B. B.; Zhou, J. *Org. Lett.* **2006**, *8*, 1283–1286.

### 3.3.5 Osmium Tetroxide Dihydroxylation

Osmium tetroxide is a highly toxic reagent. It is volatile and can cause blindness.

To a mixture of *N*-methylmorpholine-*N*-oxide·$2H_2O$ (18.2 g, 155 mmol), water (50 mL), acetone (20 mL), and osmium tetroxide (80 mg) in *t*-butanol (8 mL) was added distilled cyclohexene (10.1 mL, 100 mmol). The reaction was slightly exothermic initially and was maintained at room temperature with a water bath. The reaction was complete after stirring overnight at room temperature under nitrogen. A slurry of 1 g of sodium hydrosulfite, 12 g of magnesium silicate (magnesol), and 80 ml of water was added, and the magnesol was filtered. The filtrate was neutralized to pH 7 with 1 N $H_2SO_4$, the acetone was evaporated under vacuum, and the pH was further adjusted to pH 2. The solution was saturated with NaCl and extracted with EtOAc. The aqueous phase was concentrated by azeotroping with *n*-butanol and further extracted with ethyl acetate. The combined ethyl acetate layers were dried and evaporated, yielding 11.2 g (96.6%) of crystalline solid. Recrystallization from ether provided 10.6 g (91%) of *cis*-1,2-cyclohexanediol, mp 95–97°C.

Reference: Van Rheenen, V.; Kelly, R. C.; Cha, D. Y. *Tetrahedron Lett.* **1976**, *17*, 1973–1976.

### 3.3.6 Sharpless Asymmetric Dihydroxylation

Commercially available asymmetric dihydroxylation mixtures are AD-mix α (chiral ligand $(DHQ)_2PHAL$, $K_3Fe(CN)_6$, $K_2CO_3$, and $K_2OsO_4·2H_2O$) and AD-mix β (chiral ligand $(DHQD)_2PHAL$, $K_3Fe(CN)_6$, $K_2CO_3$, and $K_2OsO_4·2H_2O$). Reviews: (a) Noe, M. C.; Letavic, M. A.; Snow, S. L. *Org. React.* **2005**, *66*, 109–625. (b) Kolb, H. C.; VanNieuwenhze, M. S.; Sharpless, K. B. *Chem. Rev.* **1994**, *94*, 2483–2547.

To a mixture of (DHQD)$_2$PHAL (800 mg, 1.03 mmol), K$_3$Fe(CN)$_6$ (101.5 g, 308 mmol), and K$_2$CO$_3$ (42.6 g, 308 mmol) in H$_2$O:*t*-BuOH (1:1, 1000 mL) cooled to 0 °C was added K$_2$OsO$_4$(OH)$_4$ (158 mg, 0.411 mmol) followed by methanesulfonamide (9.8 g, 102.8 mmol). After stirring for 10 min at 0 °C, 3-cyclohexylacrylic acid ethyl ester (18.7 g, 102.8 mmol) was added in 1 portion. The reaction mixture was stirred at 0 °C for 18 h and then quenched with sodium sulfite (154 g). Stirring was continued for 1 h at room temperature and the solution extracted with CH$_2$Cl$_2$ (3 × 300 mL). The organic layer was washed with KOH (2 N), dried over MgSO$_4$, and evaporated to give the diol. Crystallization in heptane afforded 17.0 g (76%) of the diol as a white solid.

Reference: Alonso, M.; Santacana, F.; Rafecas, L.; Riera, A. *Org. Proc. Res. Dev.* **2005**, *9*, 690–693.

### 3.3.7 Sharpless Asymmetric Aminohydroxylation

The nitrogen sources for the aminohydroxylation are shown below.

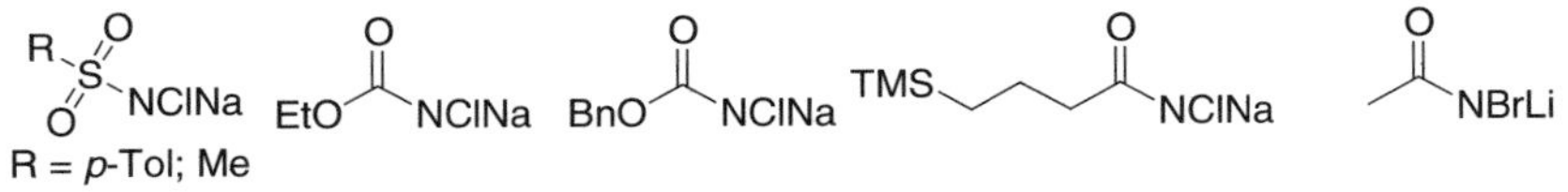

R-S(=O)$_2$-NClNa (R = *p*-Tol; Me), EtO-C(=O)-NClNa, BnO-C(=O)-NClNa, TMS-CH$_2$CH$_2$-C(=O)-NClNa, Ac-NBrLi

Sharpless, along with Noyori and Knowles, won the Nobel Prize in 2001 for his contribution to the field of asymmetric synthesis.

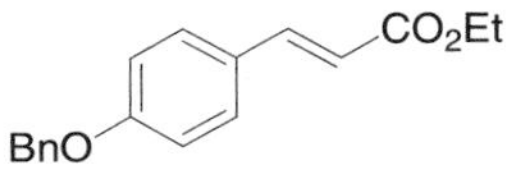

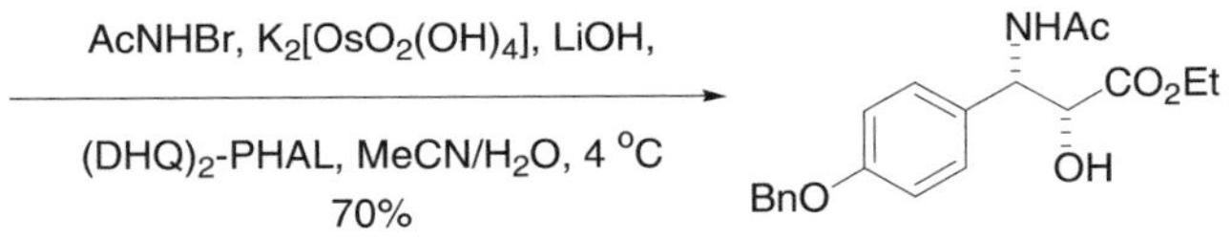

BnO-C$_6$H$_4$-CH=CH-CO$_2$Et

AcNHBr, K$_2$[OsO$_2$(OH)$_4$], LiOH, (DHQ)$_2$-PHAL, MeCN/H$_2$O, 4 °C
70%

→ BnO-C$_6$H$_4$-CH(NHAc)-CH(OH)-CO$_2$Et

In 335 mL of an aqueous solution of LiOH·H$_2$O (7.59 g, 181 mmol), K$_2$[OsO$_2$(OH)$_4$] (2.6 g, 7.1 mmol, 4 mol %) was dissolved with stirring. After addition of *t*-BuOH (665 mL), (DHQ)$_2$PHAL (6.91 g, 8.87 mmol, 5 mol %) was added, and the mixture was stirred for 10 min to give a clear solution. The solution was then diluted with additional water (665 mL) and immersed in a cooling bath set to 0 °C. A solution of the cinnamate (50.0 g, 177 mmol) in acetonitrile (335 mL) was then added to the mixture, followed by addition of *N*-bromoacetamide (26.91 g, 195.1 mmol) in one lot, and the mixture was vigorously stirred between 0 and 5 °C. After stirring for 24 h, the reaction mixture was treated with Na$_2$SO$_3$ (89 g) and stirred at room temperature for 30 min, and ethyl acetate (1 L) was added. The organic layer was separated, and the water layer was extracted three times with EtOAc. The combined organic extracts were washed with brine and dried over MgSO$_4$. After evaporation of the solvent under reduced pressure, the crude product was purified by flash chromatography on silica gel.

Elution with EtOAc:hexane (1:1) afforded 10% of the diol by-product followed by 40 g (70%) of the amino alcohol as a white crystalline solid.

Reference: Reddy, S. H. K.; Lee, S.; Datta, A.; Georg, G. I. *J. Org. Chem.* **2001**, *66*, 8211–8214.

### 3.3.8 Katsuki–Sharpless Asymmetric Epoxidation

The Katsuki–Sharpless asymmetric epoxidation is one of the first predictably enantioselective oxidations. For reviews see (a) Katsuki, T.; Martin, V. *Org. React.* **1996**, *48*, 1–299. (b) *Pfenniger, D. S. Synthesis* **1986**, 89-116.

*L*-(+)-DET = *L*-(+)-Diethyl tartrate;

TBHP = *tert*-butylhydroperoxide.

A mixture of activated 4 Å molecular sieves (1.80 g, 15–20 wt% based on geraniol) and $CH_2Cl_2$ (100 mL) was cooled to −10 °C. *L*-(+)-Diethyl tartrate (1.00 g, 4.8 mmol), titanium(IV) isopropoxide (0.91 g, 3.2 mmol), and *tert*-butylhydroperoxide (19.4 mL, 97 mmol, 5.0 M in $CH_2Cl_2$) were added sequentially. After 10 min of stirring, the mixture was cooled to −20 °C, and freshly distilled geraniol (10.0 g, 65 mmol, in 10 mL of $CH_2Cl_2$) was added dropwise over 15 min. After 45 min of stirring at −20 to −15 °C, the mixture was warmed to 0 °C. After an additional 5 min of stirring at 0 °C, the mixture was quenched sequentially with water (20 mL) and 4.5 mL of 30% aqueous NaOH saturated with solid NaCl. After 10 min of vigorous stirring, the reaction mixture was partitioned between $CH_2Cl_2$ and water. The combined organic extract was dried ($MgSO_4$) and then filtered through Celite to give a clear colorless solution. Concentration followed by bulb to bulb distillation [bp (bath) = 100 °C at 0.1 mm Hg] gave the epoxide as a colorless oil (10.3 g, 91%).

Reference: Taber, D. F.; Bui, G.; Chen, B. *J. Org. Chem.* **2001**, *66*, 3423–3426.

### 3.3.9 Shi Epoxidation

For a review on the scope of this reaction, and impact of reaction conditions on the course of the reaction, see Shi, Y. *Acc. Chem. Res.* **2004**, *37*, 488–496.

Oxone=$KHSO_4$

Aqueous Na$_2$(EDTA) (1 × 10$^{-4}$ M, 2.5 mL) and Bu$_4$NHSO$_4$ (10 mg, 0.03 mmol) were added to a solution of the olefin (77 mg, 0.5 mmol) in acetonitrile (2.5 mL) with vigorous stirring at 0 °C. A mixture of oxone (1.5 g, 2.5 mmol) and NaHCO$_3$ (0.65g, 7.75 mmol) was pulverized, and a small portion of this mixture was added to the reaction mixture to bring the pH to > 7. Then a solution of the ketone (38 mg, 0.125 mmol) in ACN (1.25 mL) was added. The rest of the Oxone and NaHCO$_3$ was added to the reaction mixture portionwise over 4.5 h. On stirring for an additional 7.5 h at 0 °C and 12 h at room temperature, the resulting mixture was diluted with water and extracted with EtOAc. The combined extracts were washed with brine, dried (Na$_2$SO$_4$), filtered, concentrated, and purified by flash chromatography to give the epoxide as a colorless oil (65 mg, 77% yield, 93% *ee* by GC, absolute stereochemistry not determined unambiguously).

Reference: Wu, X.-Y.; She, X.; Shi Y. *J. Am. Chem. Soc.* **2002**, *124*, 8792–8793.

### 3.3.10 VO(acac)$_2$, *t*-BuOOH Oxidation of Allylic Alcohols

To a solution of allylic alcohol (688 mg, 1.05 mmol) in benzene (25 mL) was added vanadyl acetylacetonate (58 mg, 0.22 mmol) and *tert*-butylhydrogenperoxide as a 1.0 M solution in toluene (2.2 mL, 2.2 mmol). The resulting blood-red slurry was stirred rapidly for 2 h, quenched with saturated sodium thiosulfate (25 mL), and the organic layer collected and dried over sodium sulfate. The solvents were removed under reduced pressure and the resulting brown residue was purified by flash chromatography (45% EtOAc/hexanes) to afford the epoxide as a 1.2:1 mixture of diastereomers (605 mg, 86%).

Reference: Shotwell, J. B.; Krygowski, E. S.; Hines, J.; Koh, B.; Huntsman, E. W. D.; Choi, H. W.; Schneekloth, J. S., Jr.; Wood, J. L.; Crews, C. M. *Org. Lett.* **2002**, *4*, 3087–3089.

## 3.4 Aldehyde to Acid Oxidation State

### 3.4.1 Baeyer–Villiger Oxidation

Reviews: (a) Mihovilovic, M. D.; Rudroff, F.; Grotzl, B. *Curr. Org. Chem.* **2004**, *8*, 1057–1069 (enantioselective). (b) Strukul, G. *Angew. Chem. Int. Ed.* 1998, *37*, 134–142 (enantioselective). (c) Krow, G. R. *Org. React.* **1993**, *43*, 251–798 (traditional).

The reaction has traditionally been run with 90% trifluoroacetic peracid or 90% H$_2$O$_2$, but the explosive nature of these reagents, as well as the availability of safer alternatives, makes their use unadvisable. The group better able to stabilize a positive charge will be the group to migrate.

To a stirred solution of the ketone (200 mg, 1.31 mmol, 1 equivalent) in $CH_2Cl_2$ (10 mL) at room temperature was added scandium triflate (32 mg, 0.066 mmol, 0.05 equivalents). After stirring for 10 min, m-chloroperoxybenzoic acid (450 mg, 2.63 mmol, 2 equivalents) was added. The reaction was stirred for 3 h and quenched with $Na_2SO_3$-doped saturated $NaHCO_3$ solution. After stirring for 10 min, the layers were separated and the combined organics were washed with saturated NaCl, dried over $MgSO_4$, vacuum filtered, and concentrated in vacuo. The crude product was purified by flash chromatography (30% EtOAc/hexanes) to yield 158 mg (71%) of the lactone as a clear oil.

Reference: Chandler, C. L.; Phillips, A. J. *Org. Lett.* **2005**, *7*, 3493–3495.

## 3.4.2 Sodium Chlorite Oxidation

A clean vessel was charged with water (20 gal), sodium hydrogen phosphate (25 kg, 183.5 mol), 5-(2-formyl-pyridin-2-yloxy)-benz[1,2,5]oxadiazole (15 kg, 62 mol), and *tert*-butyl alcohol (99 gal). The mixture was stirred for 1 h, and then a solution of sodium chlorite (34 kg, 379 mol) in water (40 gal) was added at a rate to keep the internal temperature < 35 °C. The reaction was quenched with a solution of sodium bisulfite (90 kg) in water (99 gal) at a rate to keep the temperature < 25 °C. The *tert*-butyl alcohol was stripped at slightly below atmospheric temperature (to control the bisulfite fumes released) until the head temperature was 80 °C. After cooling to 20 °C, the solids were stirred for 5.5 h, filtered, and washed with water (10 gal). The wet cake was reslurried in water (30 gal) for 1 h at 80 °C. After cooling to 20–25 °C, the solids were slurried for 2 h, filtered, and washed with water (5 gal). The solids were dried in a vacuum oven at 45–55 °C until the KF was < 0.5%, yielding 14.3 kg (89%) of the desired product.

Reference: Ruggeri, S. G.; Bill, D. R.; Bourassa, D. E.; Castaldi, M. J.; Houck, T. L.; Ripin, D. H. B.; Wei, L.; Weston, N. *Org. Process Res. Dev.* **2003**, *7*, 1043–1047.

## 3.5 Heteroatom Oxidations

### 3.5.1 Amine to Nitrone

Solution A: The amine (68.3 g, 0.35 mol) was dissolved in 1750 mL of $CH_2Cl_2$ and cooled to 5 °C (ice bath).

Solution B: *m*-CPBA (175.1 g, 0.7 mol; containing 24 wt % water and 7 wt % 3-Cl-benzoic acid) was dissolved in 1750 mL of $CH_2Cl_2$ and cooled to 5 °C (ice bath).

Solution B (except for the water) was added dropwise to solution A over a period of 45 min at 5–10 °C. The mixture was stirred for another 30 min at 10–15 °C. The oxidation was considered complete when the amine content was less than 6%. The reaction mixture was washed with 750 mL of a 10% $Na_2CO_3$ solution. The layers were separated, and the organic phase was vigorously stirred with 280 mL of 10% aqueous $Na_2SO_3$ solution for 1 h. (*Note: the aqueous layer should be treated with $Na_2SO_3$ before disposal.*) The organic layer was evaporated to half its volume, and 1050 mL of toluene was added. This mixture was heated to 80 °C while distilling off the remaining $CH_2Cl_2$. The mixture was kept at 80 °C until the reflux stopped. The resulting toluene solution (HPLC 95.5%) was processed in the cycloaddition step without purification.

Reference: Stappers, F.; Broeckx, R.; Leurs, S.; Van Den Bergh, L.; Agten, J.; Lambrechts, A.; Van den Heuvel, D.; De Smaele, D. *Org. Proc. Res. Dev.* **2002**, *6*, 911–914.

### 3.5.2 Sulfide to Sulfoxide Oxidation

A variety of reagents will accomplish this oxidation, but the stoichiometry and temperature must be carefully controlled to avoid over-oxidation (see, for example, Uemura, S. Oxidation of Sulfur, Selenium and Tellurium. In *Comprehensive Organic Synthesis*; Trost, B. M., Ed.; Pergamon Press: Oxford, 1991; Vol. 7, 758–769 and Kowalksi, P.; Mitka, K.; Ossowska, K.; Kolarska, Z. *Tetrahedron*, **2005**, *61*, 1933–1953).

To a stirred solution of the sulfide (8.11 g, 15.5 mmol) in $CH_2Cl_2$ (150 mL) was added 90% *m*-CPBA (2.97 g, 15.5 mmol) at −78 °C. The reaction mixture was stirred at −78 °C for 4 h, warmed to −20 °C, and quenched with saturated aqueous $Na_2CO_3$. The solution was washed with saturated aqueous $Na_2CO_3$, brine, and dried over anhydrous $Na_2SO_4$. Removal of solvent and recrystallization from EtOAc and hexane gave the sulfoxide (7.00 g, 13.0 mmol, 84%).

Reference: Crich, D.; Banerjee, A.; Yao, Q. *J. Am. Chem. Soc.* **2004**, *126*, 14930–14934.

### 3.5.3 Sulfide to Sulfone Oxidation

Similar conditions can be used to oxidize sulfoxides to sulfones.

To a vigorously stirred solution of the sulfide (7.47 g, 33.3 mmol) in THF (22 mL), MeOH (22 mL), and $H_2O$ (22 mL) at 0 °C was added Oxone (57.3 g, 93.2 mmol) portionwise. After 5 min at 0 °C, the white suspension was warmed to room temperature and stirred for 30 min. The reaction was poured into $H_2O$ (500 mL) and extracted with $CH_2Cl_2$ (3 × 150 mL), and the combined organic layers were dried ($Na_2SO_4$) and concentrated to give the sulfone (8.54 g, > 99% yield) as a white solid.

Reference: Voight, E. A.; Roethle, P. A.; Burke, S. D. *J. Org. Chem.* **2004**, *69*, 4534–4537.

### 3.5.4 Kagan Asymmetric Sulfur Oxidation

Considerable advances have been made recently towards the enantioselective oxidation of sulfides to sulfoxides. The reactions can be technically challenging but high enantioselectivies can be achieved.

(−)-Diethyl *D*-tartrate (12.04 g, 58.4 mmol) was dissolved in $CH_2Cl_2$ (109 mL) and the water content of the resulting clear solution determined by Karl Fischer titration. This solution was transferred to a thoroughly dried reaction vessel containing the sulfide (10.85 g, 29.2 mmol) under an inert atmosphere and stirred to give a pale yellow solution. Titanium isopropoxide (8.96 mL, 29.2 mmol) was added, followed by distilled water (0.50 mL, 27.8 mmol, 0.95 equivalents, to bring the total amount of water to 1.00 equivalents). The reaction mixture was cooled to −15 °C and cumeme hydroperoxide (5.56 mL of an 80% w/w solution, 30.7 mmol) added dropwise over 60 min, maintaining the temperature at −15 °C. After 5–16 h the reaction was complete, and a solution of 3 M HCl (60 mL, 200 mmol, 6.84 equivalents) was added, allowing the mixture to warm to 20 °C. (*Note: This addition is strongly exothermic!*) This was stirred at 20 °C for 1 h after which the pale yellow lower $CH_2Cl_2$ phase was separated from the bright orange upper aqueous phase and returned to the reaction vessel. A solution of 4 M NaOH (66 mL, 264 mmol, 9.04 equivalents) was added to the organic phase and the mixture heated to 40 °C for 1 h before cooling back to 20 °C. (*Note: This addition can be exothermic at first if there is residual HCl left in the organic phase.*) The phases were separated, and the lower $CH_2Cl_2$ phase was washed twice with water (66 mL each, 6.1 volumes) before concentration under reduced pressure to give the crude sulfoxide compound as a pale yellow oil (10.34 g, 91.4% yield, 93.6% *ee*).

Reference: Bowden, S. A.; Burke, J. N.; Gray, F.; McKown, S.; Moseley, J. D.; Moss, W. O.; Murray, P. M.; Welham, M. J.; Young, M. J. *Org. Proc. Res. Dev.* **2004**, *8*, 33–44.

# 4

# Reductions

## 4.1 Alcohols to Alkanes

### 4.1.1 Tributyltin Hydride/2,2'-Azobisisobutyronitrile (Barton Deoxygenation)

The Barton deoxygenation (or Barton–McCombie deoxygenation) is a two-step reaction sequence for the reduction of an alcohol to an alkane. The alcohol is first converted to a methyl xanthate or thioimidazoyl carbamate. Then, the xanthate or thioimidazoyl carbamate is reduced with a tin hydride reagent under radical conditions to afford the alkane. Trialkylsilanes have also been used as the hydride source. Reviews: (a) McCombie, S. W. In *Comprehensive Organic Synthesis*; Trost, B. M.; Fleming, I., Eds.; Pergamon Press: Oxford, U. K., 1991; Vol. 8, Chapter 4.2: Reduction of Saturated Alcohols and Amines to Alkanes, pp. 818–824. (b) Crich, D.; Quintero, L. *Chem. Rev.* **1989**, *89*, 1413–1432.

#### 4.1.1.1 Via a Methylxanthate

To a solution of the β-hydroxy-*N*-methyl-*O*-methylamide (0.272 g, 1.55 mol) in tetrahydrofuran (THF) (30 mL) were added carbon disulfide (6.75 mL, 112 mmol) and iodomethane (6.70 mL, 108 mmol) at 0 °C. The mixture was stirred at this temperature for 0.25 h, and then sodium hydride (60% suspension in mineral, 136.3 mg, 3.4 mmol) was added. After 20 min at 0 °C, the reaction was quenched by slow addition to 60 g of crushed ice. (Caution: hydrogen gas evolution!). The mixture was raised to room temperature and separated, and the aqueous layer was extracted with $CH_2Cl_2$ (4 × 15 mL). The combined organic extracts were dried ($Na_2SO_4$), concentrated in vacuo, and purified ($SiO_2$, 5% EtOAc in hexanes) to afford 0.354 g (86%) of the xanthate. To a solution of the xanthate (2.95 g, 11.1 mmol) in toluene (100 mL) was added tributyltin hydride (15.2 mL, 56.6 mmol) and 2,2′-azobisisobutyronitrile (AIBN, 0.109 g, 0.664 mmol). The reaction mixture was then heated to reflux for 1 h. The mixture was cooled, concentrated in vacuo, and purified ($SiO_2$, 100% hexanes to remove tin byproducts, followed by 10% EtOAc in hexanes to elute product) to afford 1.69 g (96%) of the *N*-methyl-*O*-methylamide.

Reference: Calter, M. A.; Liao, W.; Struss, J. A. *J. Org. Chem.* **2001**, *66*, 7500–7504.

### 4.1.1.2 Via a Thioimidazoyl Carbamate

A mixture of the β-hydroxymethyl ester (10.0 g, 25 mmol) and 1,1′-thiocarbonyl-imidazole (9.0 g, 50 mmol) in anhydrous THF (130 mL) was heated at reflux for 16 h. The solvent was removed under reduced pressure. The residue was dissolved in EtOAc (100 mL), and the resulting solution was washed with 0.5 N HCl (3 × 100 mL). The organic layer was dried, filtered, and concentrated. The residue was recrystallized from EtOAc/hexane to give 7.6 g (60%) of imidazolide thioester. The filtrate was concentrated, and the residue was purified by flash column chromatography using EtOAc/hexane as the eluent to give an additional 1.4 g (11%) of product.

A solution of the imidazolide thioester (5.0 g, 9.8 mmol) in dry toluene (130 mL) was treated at 100 °C with tributyltin hydride (3.4 mL, 12.6 mmol) followed by AIBN (0.1 g, 0.06 mmol), and the mixture was stirred at 100 °C for 10 min. The solvent was removed in vacuo, and the residue was dissolved in acetonitrile (100 mL) and washed with hexanes (3 × 100 mL). The acetonitrile layer was concentrated, and the residue was purified by flash chromatography using EtOAc/hexanes (0–50% mixture) as the eluent to deliver 3.6 g (95%) of the cyclopentane.

Reference: Chand, P.; Kotian, P. L.; Dehghani, A.; El-Kattan, Y.; Lin, T.-H.; Hutchison, T. L.; Babu, Y. S.; Bantia, S.; Elliott, A. J.; Montgomery, J. A. *J. Med. Chem.* **2001**, *44*, 4379–4392.

## 4.2 Aldehydes, Amides, and Nitriles to Amines

### 4.2.1 Reductive Amination with Sodium Triacetoxyborohydride

A popular and effective method for converting aldehydes to amines is through a reductive amination protocol. Typically, the aldehyde and amine react to form an intermediate imine or iminium ion and then a reducing agent (i.e., sodium triacetoxyborohydride) is added to carry out the reduction of the intermediate species to afford the amine. The reaction is accelerated in the presence of acetic acid (0.1 –3 equivalents). Reviews: (a) Baxter, E. W.; Reitz, A. B. *Org. React.* **2002**, *59*, 1–714. (b) Hutchins, R. O. In *Comprehensive Organic Synthesis*; Trost, B. M.; Fleming, I., Eds.; Pergamon Press: Oxford, U. K., 1991; Vol. 8, Chapter 1.2: Reduction of C=N to CHNH by Metal Hydrides, pp. 25–54.

Triethylamine (0.32 mL, 2.22 mmol) and H-Phe-Leu-OMe (0.72 g, 2.16 mmol) were added to a solution of the aldehyde (0.67 g, 1.66 mol) in 1,2-dichloroethane (25 mL). The solution was stirred for 5 min, after which time sodium triacetoxyborohydride (0.57 g, 2.67 mmol) was added in 1 portion. Stirring was continued for 45 min followed by addition of saturated aqueous $NaHCO_3$ and separation of the two phases. The aqueous phase was extracted with $CH_2Cl_2$. The combined organic phases were washed with brine, dried over $Na_2SO_4$, and concentrated under reduced pressure. The residue was purified by flash chromatography (heptane/EtOAc (3:1 to 1:1) to give 0.93 g (82%) of the secondary amine as a slightly yellow oil.

Reference: Blomberg, D.; Hedenström, M.; Kreye, P.; Sethson, I.; Brickmann, K.; Kihlberg, J. *J. Org. Chem.* **2004**, *69*, 3500–3508.

### 4.2.2 Lithium Aluminum Hydride Reduction of an Amide

See section 4.3.1 for a description of lithium aluminum hydride (LAH).

A mixture of the octahydroindolone (259 mg, 1.0 mmol) and LiAlH$_4$ (76 mg, 2.0 mmol) in THF (5 mL) was stirred at reflux for 3 h and then successively treated with water (1 mL), 8% sodium hydroxide (3 mL), and water (3 mL). *Note: extreme care should be practiced during the quenching process. As this process is exothermic and produces flammable hydrogen gas, it is highly advisable to cool the reaction mixture to 0 °C prior to quenching and to add the water and aqueous solution of sodium hydroxide cautiously.* The mixture was extracted with CHCl$_3$ (3 × 20 mL). The organic layers were concentrated and the crude product purified via chromatography eluting CHCl$_3$/methanol 10:1) to afford 241 mg (98%) of the octahydroindole as a pale yellow oil.

Reference: Yasuhara, T.; Nishimura, K; Yamashita, M.; Fukuyama, N.; Yamada, K.; Muraoka, O.; Tomioka, K. *Org. Lett.* **2003**, *5*, 1123–1126.

### 4.2.3 Lithium Aluminum Hydride Reduction of a Nitrile

See section 4.3.1 for a description of LAH.

A solution of LAH (1.0 M in ether, 36.5 mL, 36.5 mmol) was cooled to 0 °C, and a solution of the cyanoethyl indole (2.78 g, 16.3 mmol) was added slowly. Then, the solution was heated at reflux for 3 h. It was cooled to 0 °C and quenched by dropwise addition of water (20 mL) followed by 1 N sodium hydroxide (40 mL). The phases were separated, and the aqueous phase was extracted with ether. The combined organic phases were washed with brine and then dried (potassium hydroxide). Evaporation of the solvent gave 2.6 g (92%) of homotryptamine as a yellow oil, which solidified on standing. The hydrochloride was prepared by dissolving the amine in a minimum of ethanol and then a saturated solution of hydrogen chloride in ether was added until no additional salt formation was observed. The hydrochloride was recrystallized from ethanol.

Reference: Kuehne, M. E.; Cowen, S. D.; Xu, F.; Borman, L. S. *J. Org. Chem.* **2001**, *66*, 5303–5316.

## 4.3 Carboxylic Acids and Derivatives to Alcohols

### 4.3.1 Lithium Aluminum Hydride

LiAlH$_4$ (LAH) is a non-selective hydride reducing agent that reduces a host of functional groups, including carboxylic acids, amides, esters, lactones, ketones, aldehydes, epoxides, and nitriles, to the corresponding alcohols and amines. Great care should be exercised in quenching excess LAH. Although several quenching methods exist, an excellent way to destroy LAH and remove the aluminium salts is to carefully add water (× mL), 15% aqueous NaOH (× mL), and water (4 × mL), again in sequential order, at 0 °C to the reaction mixture. Reviews: (a) Seyden-Penne, J. *Reductions by the Alumino- and Borohydrides in Organic Synthesis*; Wiley-VCH: New York, 1997, 2nd edition. (b) Brown, H. C.; Krishnamurthy, S. *Tetrahedron* **1979**, *35*, 567–607. (c) Brown, W. G. *Org. React.* **1952**, *6*, 469–509.

#### 4.3.1.1 Reduction of a Carboxylic Acid

To a suspension of LAH (6.50 g, 171 mmol) in THF (90 mL) at 0 °C was added dropwise a solution of the diacid (3.40 g, 17.0 mmol) in THF (30 mL). The mixture was heated at reflux for 1 h and then cooled to 0 °C. The mixture was then treated dropwise with water (13 mL) and 10% aqueous sodium hydroxide (10 mL). *Note: extreme care should be practiced during the quenching process. As this process is exothermic and produces flammable hydrogen gas, it is highly advisable to cool the reaction mixture to 0 °C prior to quenching and to add the water and aqueous solution of sodium hydroxide cautiously.* The mixture was filtered over Na$_2$SO$_4$, and the filtrate was evaporated under reduced pressure to give 2.65 (91%) of the diol.

Reference: Brocksom, T. J.; Coelho, F.; Deprés, J.-P.; Greene, A. E.; Freire de Lima, M. E. Hamelin, O.; Hartmann, B.; Kanazawa, A. M.; Wang, Y. *J. Am. Chem. Soc.* **2002**, *124*, 15313–15325.

#### 4.3.1.1 Reduction of an Ester

A solution of the methyl ester (3.35 g, 14.0 mmol) in dry THF (50 mL) was added dropwise to a slurry of LAH (0.797 g, 21.0 mmol) in dry THF at 0 °C under a nitrogen atmosphere. When the addition was complete, the cooling bath was removed and the mixture was stirred at ambient temperature. After 16 h, thin layer chromatography (TLC 2:1 hexanes/EtOAc) revealed that the reaction was complete. The mixture was cooled to 0 °C, and the reaction was carefully quenched with water (0.80 mL), 15% aqueous sodium hydroxide (0.80 mL), and water (2.4 mL). *Note: extreme care should be practiced during the quenching process. As this process is exothermic and produces flammable hydrogen gas, it is highly advisable to cool the reaction mixture to 0 °C prior to quenching and to add the water and aqueous solution of sodium hydroxide cautiously.* After being stirred for 30 min, the mixture was diluted with EtOAc (100 mL), filtered through Celite (EtOAc wash), and concentrated under reduced pressure to give 3.0 g (100%) of the benzyl alcohol as a white solid.

Reference: Barbachyn, M. R.; Cleek, G. J.; Dolak, L. A.; Garmon, S. A.; Morris, J.; Seest, E. P.; Thomas, R. C.; Toops, D. S.; Watt, W.; Wishka, D. G.; Ford, C. W.; Zurenko, G. E.; Hamel, J. C.; Schaadt, R. D.; Stapert, D.; Yagi, B. H.; Adams, W. J.; Friis, J. H.; Slatter, J. G.; Sams, J. P.; Olien, N. L.; Zaya, M. J.; Wienkers, L. C.; Wynalda, M. A. *J. Med. Chem.* **2003**, *46*, 284–302.

### 4.3.2 Alane

Alane must be prepared in situ. It is an excellent reagent for the 1,2-reduction of $\alpha,\beta$-unsaturated esters. It selectively reduces esters in the presence of halogens and nitro groups. Reviews: (a) Seyden-Penne, J. *Reductions by the Alumino- and Borohydrides in Organic Synthesis*; Wiley-VCH: New York, 1997, 2$^{nd}$ edition. (b) Brown, H. C.; Krishnamurthy, S. *Tetrahedron* **1979**, *35*, 567–607.

PhS–CH(CH$_3$)–CH=CH–CO$_2$Me $\xrightarrow[\text{93\%}]{\text{AlH}_3, \text{ether, 0 °C}}$ PhS–CH(CH$_3$)–CH=CH–CH$_2$OH

To a suspension of LAH (0.60 g, 16 mmol) in dry ether (50 mL) at 0 °C was added a solution of aluminum trichloride (3.2 g, 24 mmol) in dry ether (40 mL) over a period of 10 min. The above mixture was stirred for 30 min, and the solution of the $\alpha,\beta$-unsaturated ester (3.7 g, 16 mmol) in dry ether (25 mL) was added over a period of 10 min at 0 °C and stirred at the same temperature for 1 h. The reaction mixture was diluted with ether (100 mL), and small ice pieces *were added carefully*. The solid formed was filtered, and the filtrate was evaporated to afford 3.12 g (93%) of the allylic alcohol as a viscous liquid.

Reference: Raghavan, S.; Reddy, S. R.; Tony, K. A.; Kumar, Ch. N.; Varma, A. K., Nangia, A. *J. Org. Chem.* **2002**, *67*, 5838–5841.

### 4.3.3 Borane

Borane is typically used as a THF (BH$_3$·THF) or dimethylsulfide complex (BH$_3$·SMe$_2$). Although the reactivity of the two complexes is similar, the boron dimethylsulfide species is more stable over longer periods of time. Borane will chemoselectively reduce a carboxylic acid in the presence of an ester or nitrile. Reviews: (a) Seyden-Penne, J. *Reductions by the Alumino- and Borohydrides in Organic Synthesis*; Wiley-VCH: New York, 1997, 2$^{nd}$ edition. (b) Brown, H. C.;

Krishnamurthy, S. *Tetrahedron* **1979**, *35*, 567–607. (c) Lane, C. F. *Chem. Rev.* **1976**, *76*, 773–799.

$$HO\underset{O}{\overset{O}{\|}}C-(CH_2)_6-C(=O)-O-CH_3 \xrightarrow[\text{THF, }-18\,°C \text{ to rt}]{BH_3\cdot THF} HO-(CH_2)_6-C(=O)-O-CH_3$$
88%

A 250-mL round-bottomed flask equipped with a stirring bar was charged with azelaic acid monomethyl ester (5.0 g, 24.7 mmol, 85% tech) and THF (12 mL). The solution was cooled to −18 °C (ice–salt bath) before the addition of BH$_3$·THF complex (24.7 mL of a 1.0 M solution in THF, 24.7 mmol) over 20 min. The reaction was stirred well, and the ice-bath was allowed to equilibrate slowly to room temperature. Progress of the reaction was monitored by $^{13}$C NMR. After 4 h, the reaction mixture was quenched with water (50 mL) at 0 °C and potassium carbonate (5.9 g) was added. The mixture was diluted with ethyl ether (100 mL), and the organic phase was separated. The aqueous phase was extracted with ethyl ether (3 × 100 mL), and the organic layers combined, washed with brine, and dried over anhydrous magnesium sulfate. The solvent was removed under reduced pressure, and the crude product was purified by silica gel chromatography, with 30% EtOAc/hexanes, producing a yellow oil (88% yield based on 85% pure azelaic acid monomethyl ester).

Reference: Gung, B. W.; Dickson, H. *Org. Lett.* **2002**, *4*, 2517–2519.

### 4.3.4 Lithium Borohydride

Lithium borohydride is more reactive than sodium borohydride, but less reactive than LiAlH$_4$. Reviews: (a) Seyden-Penne, J. *Reductions by the Alumino- and Borohydrides in Organic Synthesis*: Wiley-VCH; New York, 1997, 2$^{nd}$ edition. (b) Brown, H. C.; Krishnamurthy, S. *Tetrahedron* **1979**, *35*, 567–607.

#### 4.3.4.1 Reduction of a Carboxylic Acid

$$\underset{Bn}{\overset{NH_2}{\phantom{|}}}\!\!-\!\!CH(OH)\!-\!C(=O)OH \xrightarrow[\text{THF, 0 °C to rt}]{LiBH_4,\ TMSCl} \underset{Bn}{\overset{NH_2}{\phantom{|}}}\!\!-\!\!CH\!-\!CH_2OH$$
99%

To a 0 °C solution of lithium borohydride (1.32 g, 60.54 mmol) in THF (30 mL) was added trimethylsilyl chloride (15.40 mL, 121.1 mmol). The ice–water bath was removed, and the mixture was stirred at room temperature for 15 min. The mixture was cooled to 0 °C and (*S*)-phenylalanine (5.00 g, 30.27 mmol) was added. The ice–water bath was removed, and the reaction mixture was stirred for 16 h. The mixture was cooled again to 0 °C and methanol (45 mL) was added dropwise followed by a sodium hydroxide solution (2.5 M, 25 mL). The organic solvents were evaporated in vacuo and the residue was extracted with CHCl$_3$ (5 × 50 mL). The combined extracts were dried (Na$_2$SO$_4$), filtered, and evaporated in vacuo to leave 4.55 g (99%) of (*S*)-phenylalaninol as a white crystalline solid.

Reference: Organ, M. G.; Bilokin, Y. V.; Bratovanov, S. *J. Org. Chem.* **2002**, *67*, 5176–5183.

### 4.3.4.2 Reduction of a Lactone

A solution of the lactone (3.40 g, 13.18 mmol) in dry THF (10 mL) was added dropwise to a solution of lithium borohydride (2 M in THF, 16 mL, 32 mmol) in THF (80 mL) at 0 °C. The mixture was stirred overnight allowing room temperature to be reached. The reaction was poured into a cold mixture of saturated ammonium chloride (300 mL) and diethyl ether (300 mL), the phases were separated, and the aqueous layer extracted with ether. The combined organic extracts were washed with water and brine, dried (magnesium sulfate), and concentrated. The residue was purified by chromatography [150 g silica gel, hexanes/EtOAc (3:1 to 3:2)] to afford 3.391 g (98%) of the diol as a colorless oil.

Reference: Ahmed, A.; Hoegenauer, E. K.; Enev, V. S.; Hanbauer, M.; Kaehlig, H.; Öhler, E.; Mulzer, J. *J. Org. Chem.* **2003**, *68*, 3026–3042.

### 4.3.5 Sodium Borohydride/Boron Trifluoride Etherate

Sodium borohydride (5.06 g, 133 mmol) was added in portions to a solution of 2-(carboxymethyl)-4-nitrobenzoic acid (10.0 g, 44.4 mmol) in THF (220 mL). The contents were cooled to 0 °C, and boron trifluoride diethyl etherate (21.3 mL, 133 mmol) was added dropwise over 1 h. The mixture was allowed to warm to 25 °C and stirred for 16 h. The reaction mixture was then cooled to 0 °C and cautiously quenched with aqueous sodium hydroxide (1 N, 178 mL). The contents were stirred for 3 h, and the THF was removed under vacuum. The resulting aqueous suspension was cooled to 0 °C and the product was filtered off. After drying, 7.78 g (89%) of the diol was afforded as a white solid.

Reference: Quallich, G. J.; Makowski, T. W.; Sanders, A. F.; Urban, F. J.; Vazquez, E. *J. Org. Chem.* **1998**, *63*, 4116–4119.

## 4.4. Esters and Other Carboxylic Acid Derivatives to Aldehydes

### 4.4.1 Diisobutylaluminium Hydride

Diisobutylaluminium hydride (DIBAL–H) is a bulky hydride reducing agent that is very useful for the stereoselective reduction of prochiral ketones and reductions at low temperatures of the carbonyl functionality. Saturated esters are reduced to aldehydes at temperatures below −70 °C; however, α, β unsaturated esters are reduced to the allylic alcohols even with careful monitoring of the internal reaction temperature. *N*-Methyl-*O*-methyl amides (Weinreb amides) are reduced to aldehydes. In addition, a lactone can be reduced to a lactol or further reduced to the diol. Reviews (a) Seyden-Penne, J. *Reductions by the Alumino- and Borohydrides in Organic Synthesis*; Wiley-VCH: New York, 1997, 2nd edition. (b) Brown, H. C.; Krishnamurthy, S. *Tetrahedron* **1979**, *35*, 567–607. (c) Winterfeldt, E. *Synthesis* **1975**, 617–630.

DIBAL-H

### 4.4.1.1 Reduction of an Ester

To a solution of the methyl ester (3.72 g, 6.87 mmol) in CH$_2$Cl$_2$ (70 mL) at −78 °C was added DIBAL-H (8.2 mL of a 1.0 M solution in toluene, 8.2 mmol) slowly, *maintaining an internal temperature below −76 °C*. The reaction was stirred for 30 min and then quenched with methanol (3 mL). A saturated aqueous solution of Rochelle's salt (sodium potassium tartrate, 130 mL) was added, and the biphasic mixture stirred overnight. The layers were separated and the aqueous layer was extracted with CH$_2$Cl$_2$ (3 × 40 mL). The combined organic extracts were washed with brine, dried over magnesium sulfate, and concentrated. Purification of the crude product by chromatography on SiO$_2$ (5% EtOAc/hexane) gave 3.28 g (94%) of the aldehyde as a clear oil.

Reference: Dineen, T. A.; Roush, W. R. *Org. Lett.* **2004**, *6*, 2043–2046.

### 4.4.1.2 Reduction of a Weinreb Amide

To a solution of the *N*-methyl-*O*-methyl amide (1.65 g, 3.06 mmol) in THF (30 mL) at −78 °C was added DIBAL-H (2.90 mL of a 1.5 M solution in toluene, 4.35 mmol). After the reaction was stirred for 45 min, a saturated solution of potassium sodium tartrate (100 mL) was added and the mixture extracted with ether. The combined organic layers were dried, filtered, and concentrated. Purification by flash chromatography afforded 1.37 g (93%) of the aldehyde as a colorless oil.

### 4.4.1.3 Reduction of a Lactone

To a solution of the γ-lactone (0.880 g, 5.47 mmol) in dry $CH_2Cl_2$ (30 mL) maintained at −78 °C was added DIBAL-H (1.5 M in toluene, 4.0 mL, 6 mmol), and the mixture was stirred for 30 min at −78 °C. Solid ammonium chloride (0.4 g) and methanol (1 drop) were added, and the mixture was filtered through a short column of silica gel which was subsequently rinsed with 5% methanol in EtOAc. The filtrate was concentrated under reduced pressure, and the residue was chromatographed (20 g of silica gel, EtOAc:hexane, 1:1) to afford 0.621 g (76%) of the lactol as a colorless oil.

Reference: White, J. D.; Hrnciar, P. *J. Org. Chem.* **2000**, *65*, 9129–9142.

### 4.4.1.4 Reduction of a Nitrile

A flame-dried, 25-mL, round-bottomed flask was charged with the nitrile (0.150 g, 0.510 mmol) and dry toluene (6.5 mL). The solution was cooled to −78 °C under a nitrogen atmosphere and DIBAL-H (1.5 M in toluene, 0.68 mL, 1.02 mmol) was added slowly. The resulting reddish-orange solution was allowed to warm to ambient temperature over 1 h at which time 0.2 mL of acetone, 0.2 mL of EtOAc, and 0.2 mL pH 7 phosphate buffer were added in sequence. The mixture was then stirred vigorously for 20 min and anhydrous $Na_2SO_4$ was added, maintaining the vigorous stirring for an additional 20 min. The resulting yellow solution was filtered over a pad of silica gel and $Na_2SO_4$. The filtrate was concentrated, and the residue thus afforded was purified via radial chromatography (20% EtOAc/hexanes) to afford 0.140 g (92%) of the aldehyde as a light yellow viscous oil.

Reference: Andrus, M. B.; Meredith, E. L.; Hicken, E. J.; Simmons, B. L.; Glancey, R. R.; Ma, W. *J. Org. Chem.* **2003**, *68*, 8162–8169.

## 4.4.2 Triethylsilane and Pd/C (Fukuyama Reduction)

Fukuyama reduction is a mild method for the conversion of thioesters to aldehydes in the presence of other susceptible functional groups, including amides, esters, lactones, and acetonides. Review: Fukuyama, T.; Tokuyama, H. *Aldrichimica Acta* **2004**, *37*, 87–96.

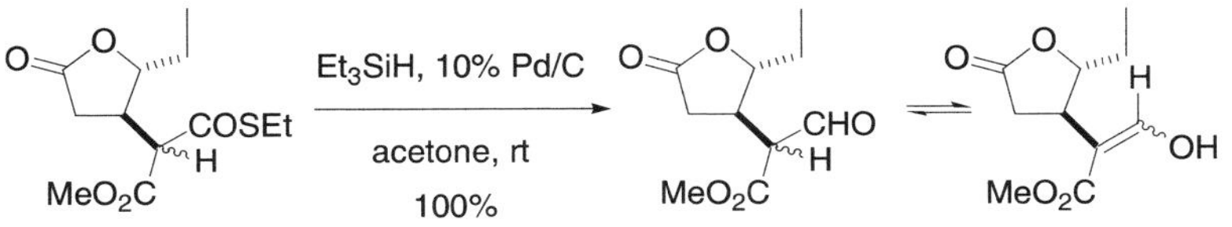

To a stirred mixture of the thioester (800 mg, 2.92 mmol) and 10% Pd on carbon (320 mg, 0.029 mmol) in acetone (7.5 mL) was added triethylsilane (1.4 mL, 8.77 mmol) at room temperature under an argon atmosphere. Stirring was continued for 3 h at room temperature. The catalyst was filtered off using Celite, and the filtrate was evaporated to give a crude residue that was chromatographed on silica gel (EtOAc/n-hexane, 1:2 to 1:1) to give 630 mg (100%) of an equilibrium mixture of the aldehyde and enol as a colorless oil.

Reference: Takayama, H.; Fujiwara, R.; Kasai, Y.; Kitajima, M.; Aimi, N. *Org. Lett.* **2003**, *5*, 2967–2970.

## 4.5 Ketones or Aldehydes to Alcohols

### 4.5.1 Lithium Aluminum Hydride

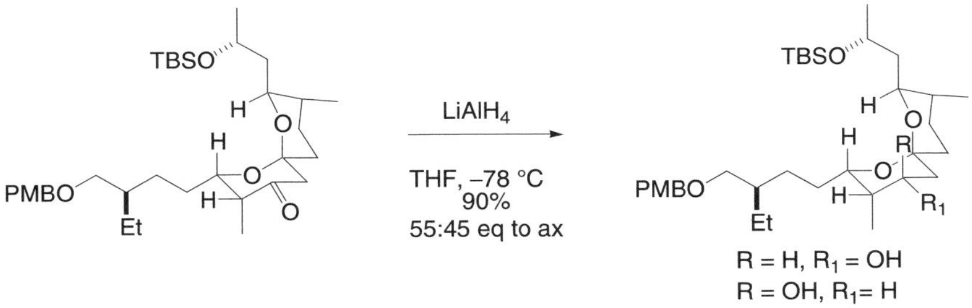

To a solution of LAH (1.0 M in THF, 1.01 mL, and 1.01 mmol) was added a solution of the ketone (596 mg, 1.01 mmol) in THF (6 mL) at −78 °C under nitrogen. After 1 h, saturated aqueous sodium potassium tartrate (6 mL) *was added carefully*, and the mixture was stirred rapidly at ambient temperature for 1 h. The resulting mixture was diluted with water (6 mL) and extracted with ether (3 × 25 mL). The combined organic extracts were dried (MgSO$_4$) and purified by flash chromatography (20% EtOAc in hexane, then 50% EtOAc in hexane) to give 330.5 mg (55%) of the equatorial alcohol and 266.1 mg (45%) of the axial alcohol as colorless oils.

Reference: Evans, D. A.; Rieger, D. L.; Jones, T. K.; Kaldor, S. W. *J. Org. Chem.* **1990**, *55*, 6260–6268.

*Note: compare diastereoselectivity of the above reaction with example 4.5.8.*

### 4.5.2 Sodium Borohydride

Sodium borohydride is a widely used mild and selective reducing agent. It selectively reduces an aldehyde or ketone in the presence of esters, lactones, carboxylic acids, and amides in methanol or THF at room temperature. Reviews: (a) Seyden-Penne, J.

*Reductions by the Alumino- and Borohydrides in Organic Synthesis*; Wiley-VCH: New York, 1997, 2nd edition. (b) Brown, H. C.; Krishnamurthy, S. *Tetrahedron* **1979**, *35*, 567–607.

A mixture of the amido ketone (1.70 g, 3.86 mmol) and sodium borohydride (293 mg, 7.72 mmol) in anhydrous methanol (60 mL) was stirred at −10 °C for 25 min. Saturated $NaHCO_3$ (40 mL) and $CH_2Cl_2$ (80 mL) were added, and the mixture was stirred at 0 °C for 5 min. The organic layer was removed, and the aqueous layer was extracted with $CH_2Cl_2$ (3 × 40 mL). The combined organic layers were dried ($Na_2SO_4$) and concentrated to give 1.68 g (95%) of the secondary alcohol as a white solid.

Reference: Deiters, A.; Chen, K.; Eary, C. T.; Martin, S. F. *J. Am. Chem. Soc.* **2003**, *125*, 4541–4550.

### 4.5.3 Sodium Borohydride/Cerium Trichloride (Luche Reduction)

The Luche reduction is an excellent method for the 1,2-reduction of α,β-unsaturated ketones. In addition, ketones are reduced selectively in the presence of aldehydes.

A 100-mL flask was charged with the enone (555 mg, 1.32 mmol), a magnetic stir bar, and $CH_2Cl_2$ (6 mL). The mixture was cooled to −78 °C in a dry ice–acetone bath under a nitrogen atmosphere. To this solution was added a 0.4 M solution of cerium trichloride in methanol (10 mL). After additional stirring and continued cooling, sodium borohydride (75 mg, 1.98 mmol) was added. The reaction was stirred for 2 h at −78 °C until only alcohol was visible by TLC. The reaction was diluted with ether (15 mL) and quenched at −78 °C by the addition of a 1 M aqueous solution of sodium bisulfate (4 mL). The mixture was stirred for 20 min and the phases were separated. The aqueous layer was extracted with ether (4 × 10 mL) and the combined organic layers were washed with brine and dried over $Na_2SO_4$, filtered, and concentrated under reduced pressure. Flash chromatography provided 477 mg (86%) of an anomeric mixture of the alcohol as a colorless oil.

Reference: Haukaas, M. H.; O'Doherty, G. A. *Org. Lett.* **2001**, *3*, 401–404.

### 4.5.4 Zinc Borohydride

Because $Zn^{+2}$ is a good chelating cation, highly diastereoselectivity reductions of α or β-hydroxy ketones and esters can be achieved with zinc borohydride. Reviews: (a) Narasimhan, S.; Balakumar, R. *Aldrichimica Acta* **1998**, *31*, 19–26.

(b) Seyden-Penne, J. *Reductions by the Alumino- and Borohydrides in Organic Synthesis*; Wiley-VCH: New York, 1997, 2<sup>nd</sup> edition.

$$\text{t-BuO-C(O)-CH}_2\text{-C(O)-CH}_2\text{-CH(OH)-CH}_2\text{-C(=CH}_2\text{)CH}_3 \xrightarrow[\text{CH}_2\text{Cl}_2, \text{THF}, -78°\text{C}]{\text{Zn(BH}_4)_2} \text{t-BuO-C(O)-CH}_2\text{-CH(OH)-CH}_2\text{-CH(OH)-CH}_2\text{-C(=CH}_2\text{)CH}_3$$

80%, *dr* >15:1

To a 500-mL round-bottomed flask charged with a magnetic stir bar was added keto ester (1.91 g, 7.88 mmol) and CH$_2$Cl$_2$ (158 mL). The mixture was cooled to −78 °C and a freshly prepared solution of Zn(BH$_4$)$_2$ (60 mL of 0.2 M in THF, 12 mmol) was added dropwise with stirring. The reaction was stirred for 2 h at this temperature before being quenched by the addition of a saturated aqueous solution of ammonium chloride (approximately 100 mL). The resulting solution was allowed to warm to room temperature with stirring for 12 h. The organic layer was separated, and the aqueous layer was extracted with EtOAc (three times). The combined organic phases were dried (MgSO$_4$), filtered, and concentrated. The crude oil was purified by flash chromatography (SiO$_2$, 30% to 50% EtOAc/hexanes) to give 1.54 g (80%, dr >15:1) of the diol as a colorless oil.

Reference: Dakin, L. A.; Panek, J. S. *Org. Lett.* **2003**, *5*, 3995–3998.

### 4.5.5 Diisobutylaluminum Hydride

See section 4.4.1 for a description of DIBAL–H.

$$\text{HC≡C-C(O)-CH(CH}_3\text{)-CH(OH)-CH=CH-CH}_2\text{-OTBDPS} \xrightarrow[\text{THF}, -78°\text{C}]{\text{DIBAL–H}} \text{HC≡C-CH(OH)-CH(CH}_3\text{)-CH(OH)-CH=CH-CH}_2\text{-OTBDPS}$$

98%

To solution of the ynone (0.893 g, 2.20 mmol) in THF (15 mL) at −78 °C under argon was added DIBAL-H (4.40 mmol of 1.0 M solution in CH$_2$Cl$_2$, 4.40 mmol) dropwise. After 30 min, the reaction was quenched by the addition of EtOAc (0.1 mL) and saturated aqueous sodium potassium tartrate (15 mL), and the slurry was warmed to room temperature with vigorous stirring for 12 h. The resulting clear biphase was extracted with ether (3 × 30 mL), and the combined organic layers were washed with brine (30 mL) and dried over Na$_2$SO$_4$. Evaporation in vacuo gave a pale yellow oil that was purified by silica gel chromatography (2:1 hexanes/EtOAc) to give 0.880 g (98%) of the diol.

Reference: Evans, D. A.; Starr, J. T. *J. Am. Chem. Soc.* **2003**, *125*, 13531–13540.

### 4.5.6 Lithium tri-*tert*-Butoxyaluminohydride

Lithium tri-*tert*-Butoxyaluminohydride is a bulky chemo- and stereoselective hydride reducing agent. Aldehydes are reduced chemoselectively in the presence of ketones and esters at low temperature. Ethers acetals, epoxides, chlorides, bromides, and nitro compounds are unaffected by this reagent. Reviews: (a) Seyden-Penne, J. *Reductions by the Alumino- and Borohydrides in Organic Synthesis*; Wiley-VCH: New York, 1997, 2<sup>nd</sup> edition. (b) Málek, J. *Org. React.* **1985**, *34*, 1–317.

To a solution of the ketone (20.4 g, 46.9 mmol) in THF (500 mL) at 0 °C was added lithium tri-*tert*-butoxyaluminohydride [LiAlH(O*t*-Bu)$_3$] (61.0 mL of a 1.0 M solution in THF, 61.0 mmol). The resulting reaction mixture was stirred for 30 min at 0 °C and then for 30 min at 25 °C. After the reaction was complete, as established by TLC analysis, it was quenched by the addition of saturated ammonium chloride (200 mL) followed by addition of EtOAc (300 mL). The mixture was stirred at 25 °C for 2 h, followed by extraction with EtOAc (3 × 300 mL). The combined extracts were washed with brine, dried over magnesium sulfate, and concentrated to give the crude product, which was purified by flash chromatography (silica gel, 25% EtOAc in hexanes) to afford 19.4 g (95%) of the alcohol as a white solid.

Reference: Nicolaou, K. C.; Pfefferkorn, J. A.; Kim, S.; Wei, H. X. *J. Am. Chem. Soc.* **1999**, *121*, 4724–4725.

### 4.5.7 L-Selectride

*L*-Selectride is a bulky stereoselective hydride reducing agent for prochiral ketones. Review: Seyden-Penne, J. *Reductions by the Alumino- and Borohydrides in Organic Synthesis*; Wiley-VCH: New York, 1997, 2$^{nd}$ edition.

| Other reducing agents | yield and diastereomeric ratio |
|---|---|
| NaBH$_4$, CeCl$_3$ | 90% (1:1.8) |
| DIBAL-H | 89% (1:1.1) |
| BH$_3$·SMe, oxazaborolidine | 84% (2.3:1) |
| LiAlH$_4$, chirald | 87% (5.2:1) |

To a solution of the enone (101 mg, 0.23 mmol) in dry THF (10 mL) was added *L*-selectride (0.46 mL of 1.0 M solution in THF, 0.46 mmol) dropwise at 0 °C. The reaction mixture was stirred for 0.5 h at 0 °C and then allowed to warm to room temperature for another 0.5 h. The mixture was then diluted with EtOAc (100 mL) and filtered through a pad of silica gel, which was rinsed with EtOAc (100 mL). The filtrate was concentrated under reduced pressure. The residue was purified by chromatography (hexane/EtOAc, 3:1) to give 86 mg (85%) of the *anti*-alcohol and 4.8 mg (4.7%) of the *syn*-alcohol as colorless oils in a diastereomeric ratio of 18:1.

Reference: Chun, J.; Byun, H.-S.; Arthur, G.; Bittman, R. *J. Org. Chem.* **2003**, *68*, 355–359.

### 4.5.8 Samarium Iodide/Isopropanol (Meerwein–Pondorf–Verlag)

This is a modification of the venerable Meerwin–Pondorf–Verlag reaction. Compare the example below with the reaction in section 4.5.1. Reviews: (a) de Graauw, C. F.; Peters, J. A.; van Bekkum, H.; Huskens, J. *Synthesis* **1994**, *10*, 1007–1017. (b) Kellogg, R. M. In *Comprehensive Organic Synthesis*; Trost, B. M.; Fleming, I., Eds.; Pergamon Press: Oxford, U. K., 1991; Vol. 8, Chapter 1.3: Reduction of C=X to CHXN by Hydride Delivery from Carbon, pp. 88–91.(c) Wilds, A. L. *Org. React.* **1944**, *2*, 178–223.

To a solution of the ketone (10.73 g, 18.2 mmol) in THF (62 mL) at ambient temperature under nitrogen were added degassed isopropanol (13.9 mL, 182 mmol) and samarium iodide (0.1 M in THF, 2.73 mL, 0.273 mmol). The resulting deep-green solution was stirred at ambient temperature for 18 h. Saturated $NaHCO_3$ (60 mL) was added to quench the reaction. The mixture was diluted with diethyl ether (120 mL) and the layers were separated. The organic layer was washed with saturated $NaHCO_3$ (60 mL). The combined aqueous layers were extracted with diethyl ether (2 × 60 mL). The combined organic extracts were washed with aqueous sodium sulfite (1.5 N, 60 mL) and brine (60 mL). After drying over magnesium sulfate, the organic layer was purified by flash chromatography (18% EtOAc in hexane, then 50% EtOAc in hexane) to give 10.56 g (98%) of the equatorial alcohol and 0.14 g (1%) of the axial alcohol as colorless oils.

Reference: Evans, D. A.; Rieger, D. L.; Jones, T. K.; Kaldor, S. W. *J. Org. Chem.* **1990**, *55*, 6260–6268.

### 4.5.9 Tetramethylammonium Triacetoxyborane

Tetramethylammonium triacetoxyborane is often used for the stereoselective reduction of β-hydroxyketones to form *anti*-1,3-diols.

Acetic acid (2 mL) was added to a stirred suspension of tetramethylammonium triacetoxyborohydride (1.54 g, 5.85 mmol) in acetonitrile (2 mL) at room temperature. The mixture was stirred at room temperature for 30 min and then cooled to −40 °C. The β-hydroxyketone (300 mg, 0.731 mmol) in acetonitrile (2 mL) was added dropwise at this temperature. A solution of camphorsulfonic acid (85 mg, 0.366 mmol) in a mixture of acetic acid:acetonitrile (1:1, 4 mL) was added, and the mixture was allowed to warm to −22 °C over 18 h. The mixture was poured into saturated aqueous $NaHCO_3$ (50 mL). A saturated aqueous solution of sodium potassium tartrate (50 mL) was added, followed by ether (100 mL), and the mixture was stirred vigorously at room temperature for 8 h. The layers were separated, and the aqueous layer was extracted with EtOAc (3 × 50 mL). The combined organic extracts were washed with water (50 mL), brine (50 mL), and dried ($MgSO_4$). The solvent was removed under vacuum to afford 300 mg (99.5%, dr >97:3) of the *anti*-1,3-diol as a colorless oil.

Reference: Paterson, I; Delgado, O.; Florence, G. J.; Lyothier, I.; O'Brien, M; Scott, J. P.; Sereinig, N. *J. Org. Chem.* **2005**, *70*, 150–160.

### 4.5.10 Corey–Bakshi–Shibata Reduction

The Corey–Bakshi–Shibata reduction (CBS reduction) is a highly enantioselective method for arylketones, diaryl ketones, and dialkylketones. In addition, cyclic α,β-unsaturated ketones, acyclic α,β-unsaturated ketones, and α,β-ynones are reduced enantioselectively in a 1,2-fashion. The high enantioselective nature of this reduction relies on the chiral oxazaborolidine catalyst, shown in the reaction scheme, in the presence of borane or a dialkylborane. Reviews: (a) Singh, V. K. *Synthesis* **1992**, 605–617. (b) Deloux, L.; Srebnik M. *Chem. Rev.* **1993**, *93*, 763–784. (c) Corey, E. J.; Helal, C. J. *Angew. Chem. Int. Ed.* **1998**, *37*, 1986–2012.

To a 100-mL round-bottomed flask charged with a magnetic stir bar was added a solution of (*S*)-CBS (1.0 M in toluene, 0.297 mL, 0.297 mmol). The toluene was then removed by placing the flask on a high vacuum pump for approximately 2 h. The flask was removed and THF (15 mL) was added. The mixture was cooled to −20 °C and a solution of boron dimethylsulfide complex (10 M, 0.594 mL, 5.94 mmol) was added. To this reaction mixture, a solution of the ketone (1.51 g, 2.9 mmol) in THF (15 mL) was added dropwise. The reaction was stirred for 8 h at −20 °C before methanol (15 mL) was carefully added to destroy excess borane. The reaction was diluted with saturated aqueous sodium chloride (approximately 25 mL) and extracted with EtOAc three times. The combined organic extracts were washed with brine, dried with magnesium sulfate, and concentrated. The crude oil was purified by flash chromatography ($SiO_2$, 0 to 10% EtOAc in hexanes) giving 1.21 g (80%, dr > 15:1) of the allylic alcohol as a colorless oil.

Reference: Dakin, L. A.; Panek, J. S. *Org. Lett.* **2003**, *5*, 3995–3998.

## 4.5.11 R-Alpine Borane (Midland Reduction)

Review: Midland, M. M. *Chem. Rev.* **1989**, *89*, 1553–1561.

R-Alpine-borane    S-Alpine-borane

$C_{16}H_{33}$-C(=O)-C≡C-CH$_2$CH$_2$-CO$_2$Me → (R-Alpine borane, THF, 85%, 89% ee) → $C_{16}H_{33}$-CH(OH)-C≡C-CH$_2$CH$_2$-CO$_2$Me

A solution of *R*-Alpine borane (0.5 M in THF, 57.2 mL, 28.6 mmol) was slowly added to the ynone (5.21 g, 14.3 mmol). The resulting solution was stirred for 36 h. After removal of solvent in vacuo, the residue was diluted with ether (200 mL). Ethanolamine (1.72 mL, 28.6 mmol) was slowly added, producing a yellow precipitate which was removed by filtration through Celite. The filtrate was concentrated and directly subjected to flash chromatography to yield 4.44 g (85%) of the alcohol as a white solid.

Reference: Dussault, P. H.; Eary, C. T.; Woller, K. R. *J. Org. Chem.* **1999**, *64*, 1789–1797.

## 4.5.12 Baker's Yeast

Baker's yeast offers the synthetic chemist an enzymatic method for the asymmetric reduction of β-keto esters, α-hydroxyaldehydes and ketones, and β-diketones. Reviews: (a) Rene Csuk, R.; Glaenzer, B. I. *Chem. Rev.* **1991**, *91*, 49–97. (b) Servi, S. *Synthesis*, **1990**, 1–25.

baker's yeast, sucrose
H$_2$O, 32–33 °C
84%
enantiomeric ratio = 95:5

To a 6-L Erlenmeyer flask immersed in a water bath were added, in succession, the keto- ester (34.15 g, 0.133 mol), sucrose (513 g), and distilled water (2731 mL). The reaction mixture was stirred until the sucrose dissolved, after which dry baker's yeast (341 g, Red Star™ was added. The reaction mixture was stirred at 32–33 °C for 24 h and then centrifuged. The aqueous layer was extracted with ether (four times) and the combined organic extracts dried over anhydrous Na$_2$SO$_4$ and concentrated in vacuo. The crude product was purified by flash chromatography (silica; 4:1 then 2:1 hexanes/EtOAc) to afford 28.7 g (84%) of the β-hydroxy ester as a colorless oil.

Reference: Williams, R. M; Cao, J. Tsujishima, H; Cox, R. J. *J. Am. Chem. Soc.* **2003**, *125*, 12172–12178.

### 4.5.13 2,2'-Bis(diphenylphosphino)-1,1'-binaphthyl /Hydrogen (Noyori Asymmetric Reduction)

2,2´-Bis(diphenylphosphino)-1,1´-binaphthyl (BINAP) is an excellent catalyst for the reduction of ketones under homogeneous hydrogenation conditions. Reviews: (a) Noyori, R.; Ohkuma, T. *Angew. Chem. Int. Ed.* **2001**, *40*, 40–73. (b) Singh, V. K. *Synthesis* **1992**, 605–617. (c) Noyori, R.; Takaya, H. *Acc. Chem. Res.* **1990**, *23*, 345–350.

Dry degassed DMF (3 mL) was added to a flask containing benzeneruthenium(II) chloride dimer (45 mg, 0.09 mmol) and (*R*)-(+)-BINAP (112 mg, 0.18 mmol). The slurry was heated to 90 °C with stirring for 20 min. The reddish-brown solution was allowed to cool to room temperature and was then added via cannula to a Parr flask containing a degassed solution of the β-keto ester (8.4 g, 30.0 mmol) in dry ethanol (15 mL). The hydrogenation flask was flushed several times with hydrogen and then pressurized with 4.0 bar hydrogen at 90 °C with vigorous shaking for 20 h. After the solution cooled to room temperature, the dark-red solution was concentrated in vacuo and purified by flash chromatography to provide 7.1 g (84%, >90% *ee*) of the β-hydroxy ester as a pale yellow liquid.

Reference: Herb, C.; Maier, M.E. *J. Org. Chem.* **2003**, *68*, 8129–8135.

## 4.6 Ketones to Alkanes or Alkenes

### 4.6.1 Wolff–Kishner Reduction

The Wolff–Kishner reaction is a classical reduction method of ketones to alkanes. This method involves converting the ketone to its corresponding hydrazine; treatment of the resulting hydrazine with base affords the alkane and nitrogen gas. Because elevated temperatures (> 200 °C) and strong base are required, this method is harsh when sensitive functionality is present. However, milder methods have been developed including the Myers' modification where the reaction temperatures are much lower (≤ 100 °C).

Reviews: (a) Hutchins, R.O.; Hutchins, M. K. In *Comprehensive Organic Synthesis*; Trost, B. M.; Fleming, I., Eds.; Pergamon Press: Oxford, U. K., 1991; Vol. 8, Chapter 1.14: Reduction of C=X to CH$_2$ by Wolff-Kishner and Other Hydrazone Methods, pp. 327–362. (b) Reusch, W. In *Reduction*; Austine, R. L., Ed.; Dekker: New York, 1968; pp. 171–221. (c) Szmant, H. H. *Angew. Chem. Int. Ed.* **1968**, *7*, 120–128. (d) Todd, D. *Org. React.* **1948**, *4*, 378–422.

### 4.6.1.1 Traditional

A mixture of the indolone (1.04 g, 4.2 mmol) and hydrazine monohydrate (1.12 mL, 22.4 mmol) in diethylene glycol (20 mL) was stirred at 80 °C for 1 h and then refluxed for 1 h. The resulting mixture was cooled to room temperature, treated with a solution of KOH (1.2 g, 21.4 mmol) in water (5 mL), and refluxed for 2 h. The resulting mixture was poured into water (100 mL), and the precipitate was filtered off, washed with water (5 × 50 mL), and dried to deliver 0.84 g (86%) of the indole as a greenish solid.

Reference: Kashulin, I. A.; Nifant'ev, I. E. *J. Org. Chem.* **2004**, *69*, 5476–5479.

### 4.6.1.2 Myers' Modification

A 25-mL, pear-shaped flask equipped with a Teflon-coated stir bar was charged with scandium trifluoromethanesulfonate (0.002 g, 0.0058 mmol) and hecogenin (0.250 g, 0.581 mmol), a septum cap was affixed, and a needle connected to an argon balloon was inserted through the septum cap. 1,2-Bis-(*tert*-butyldimethylsilyl)hydrazine (0.461 g, 1.28 mmol) and dry CHCl$_3$ (1.5 mL) were then introduced sequentially by syringe. The reaction flask was immersed in an oil bath, and the bath was heated to 55 °C. After stirring at 55 °C for 20 h, the heating bath was removed and the reaction solution, a colorless liquid, was allowed to cool to ambient temperature. Dry *n*-hexane (3 mL) was added, providing a white suspension. The septum cap was removed, and the reaction mixture was filtered through a plug of glass wool into a 25-mL pear-shaped receiving flask. The filtration was quantitated with *n*-hexane (2 × 1 mL). The solvents were removed in vacuo on a rotary evaporator, and then the flask was flushed with dry argon, charged with a Teflon-coated stir bar, and capped with a rubber septum. A needle affixed to an argon balloon was inserted through the septum cap. A separate 25-mL pear-shaped flask equipped with a Teflon-coated stir bar was charged with potassium *tert*-butoxide (1.01 g, 9.00 mmol), and a needle affixed to an argon balloon was inserted through the septum cap. Dry dimethylsulfoxide (7.5 mL) was added via the syringe, and the mixture was stirred at 23 °C until all particulates had dissolved (about 15 min). *tert*-Butanol (0.861 mL, 9.00 mmol) was then added via the syringe, and the resulting solution was transferred by syringe to the flask containing the white solid TBSH derivative. The reaction flask was immersed in an oil bath, and the bath was heated to 100 °C. After stirring at 100 °C for 24 h, the bath was removed and the resulting brown slurry was allowed to cool to 23 °C. The reaction mixture was then poured into a solution of half-saturated brine (30 mL, precooled to 0 °C), and the transfer was quantitated with washes of water (2 × 1 mL) and $CH_2Cl_2$ (2 × 1 mL). The resulting mixture was extracted with $CH_2Cl_2$ (7 × 4 mL). The organic extracts were combined and then dried over $Na_2SO_4$. Solids were removed by filtration and solvents were removed in vacuo. The concentrate was purified by flash chromatography (5:40:55, methanol/$CH_2Cl_2$/*n*-hexane) to provide 0.232 g (96%) of tigogenin as a white solid.

Reference: Furrow, M. E.; Myers, A. G. *J. Am. Chem. Soc.* **2004**, *126*, 5436–5445.

### 4.6.2 Desulfurization with Raney Nickel

Raney Nickel (nickel–aluminium alloy) is an effective catalyst for the conversion of thioketals to alkanes. Ultimately, a ketone is converted to the alkane with this mild method through a thioketal. Review: Pettit, G. R.; van Tamelen, E. E. *Org. React.* **1962**, *12*, 356–529.

To a solution containing the ketone (0.2 g, 0.6 mmol) in dry $CH_2Cl_2$ (20 mL) were added 1,2-ethanedithiol (0.3 g, 3.0 mmol) and boron trifluoride etherate (0.3 g, 2.3 mmol). The mixture was stirred at room temperature for 24 h and quenched with aqueous $NaHCO_3$ (10 mL). The aqueous layer was extracted with $CH_2Cl_2$, and the combined organic layers were dried over magnesium sulfate and concentrated under reduced pressure. The crude residue was purified by flash silica gel chromatography to give 0.2 g (85%) of the dithiane as a white solid.

To a solution containing the dithiane (0.6 g, 1.5 mmol) in absolute ethanol (100 mL) was added Raney nickel (2 g). The suspension was heated at reflux for 4 h, cooled to room temperature, and filtered through Celite with ethanol (200 mL). The solid was recrystallized from EtOAc to give 0.46 g (93%) of the reduced product as a white solid.

Reference: Padwa, A.; Brodney, M. A.; Lynch, S. M. *J. Org. Chem.* **2001**, *66*, 1716–1724.

### 4.6.3 Reduction of Tosylhydrazones with Sodium Cyanoborohydride

A mild method for converting ketones to alkanes is via a two-step process involving a tosylhydrazone. First, the ketone is converted to the corresponding tosylhydrazone, and then treatment of the tosylhydrazone with the mild reducing agent sodium cyanoborohydride, under slightly acidic conditions, affords the alkane. Reviews: (a) Hutchins, R. O.; Hutchins, M. K. In *Comprehensive Organic Synthesis*; Trost, B. M.; Fleming, I., Eds.; Pergamon Press: Oxford, U. K., 1991; Vol. 8, Chapter 1.14: Reduction of C=X to $CH_2$ by Wolff-Kishner and Other Hydrazone Methods, pp. 343–362. (b) Lane, C. F. *Synthesis* **1975**, 135–146.

The hydroxyketone (748.3 mg, 2.79 mmol) was dissolved in methanol (36 mL), and $TsNHNH_2$ (1.17g, 6.98 mmol) was added. The reaction was stirred for 3 h at room temperature and then concentrated and filtered through a plug of silica gel with 10% ethanol/hexanes to remove excess $TsNHNH_2$. The filtrate was concentrated, and the residue (tosylhydrazone) was dissolved in THF (25 mL) and methanol (25 mL). A trace of methyl orange indicator was added, and the solution was cooled to 0 °C. Sodium cyanoborohydride (119.2 mg, 1.90 mmol) in methanol (2 mL) was added by syringe. Aqueous hydrochloric acid (1 N, 5 × 300 μL portions at 20 min intervals) was added dropwise by syringe. The progress of the reaction was monitored by TLC (40% EtOAc/hexanes). The color of the reaction was maintained yellow–orange (pH > 3.8). After the fifth addition of 1 N HCl, all of the remaining starting material was consumed, as judged by TLC. The reaction was diluted with EtOAc (250 mL) and washed with water (50 mL), saturated $NaHCO_3$ (50 mL), and brine (50 mL). The organic layer was dried over $Na_2SO_4$ and filtered through a plug of silica gel with 100% EtOAc. The filtrate was concentrated to give a white foam (tosylhydrazine). The foam was dissolved in ethanol (52 mL), and the solution was degassed with nitrogen for 5 min. NaOAc·$H_2O$ (10.4 g) was added, and the mixture was degassed for a further 5 min. The reaction was then heated at 75 °C for 20 min, and the evolution of gas was observed. The reaction was cooled to room temperature, diluted with EtOAc (250 mL), and washed with water (50 mL) and brine (50 mL). The organic layer was dried over $Na_2SO_4$, filtered, and concentrated in vacuo. Purification by flash chromatography gave 570.5 mg (80%) of the alcohol.

Reference: Thompson, C. F.; Jamison, T. F.; Jacobsen, E. N. *J. Am. Chem. Soc.* **2001**, *123*, 9974–9983.

### 4.6.4 Zinc Amalgam (Clemmensen Reduction)

The Clemmensen reduction is a classical method for the reduction of ketones to alkanes.

Reviews: (a) Motherwell, W. B.; Nutley, C. J. *Cont. Org. Synth.* **1992**, 219–241. (b) Yamamura, S.; Nishiyama, S. In *Comprehensive Organic Synthesis*; Trost, B. M.; Fleming, I., Eds.; Pergamon Press: Oxford, U. K., 1991; Vol. 8, Chapter 1.13: Reduction of C=X to $CH_2$ by Dissolving Metals and Related Methods, pp. 309–313. (c) Vedejs, E. *Org. React.* **1975**, *22*, 401–422. (d) Martin, E. L. *Org. React.* **1942**, *1*, 155–209.

A mixture of mossy zinc (6 g) and 5% $HgCl_2$ solution (12 mL) was kept at room temperature for 1 h with occasional shaking, and the aqueous layer was then decanted. The amalgamated zinc was covered with concentrated hydrochloric acid (30 mL) and water (20 mL), and the ketone (0.8 g, 2.2 mmol) and toluene (7 mL) were added. The mixture was heated under reflux for 2.5 h, cooled, and worked up to give 0.69 g of a crude mixture. The mixture was completely hydrolyzed by refluxing for 2 h with a solution of 0.5 g of potassium hydroxide in methanol (10 mL). The methanol was removed, and the residue was diluted with water. The clear alkaline solution was acidified and worked up to furnish 0.575 g of the crude phenol. The phenol was recrystallized from ether/petroleum ether to afford 0.51 g (76%) of the phenol.

Reference: Ghosh, A. C.; Hazra, B. G.; Duax, W. L. *J. Org. Chem.* **1977**, *42*, 3091–3094.

## 4.6.5 Ionic Hydrogenation with Triethyl Silane and Boron Trifluoride

Diisobutylaluminium hydride (DIBAL–H) (1.0 M in heptane, 2.70 mL, 2.70 mmol) was added to a solution of the lactone (598 mg, 2.28 mmol) in CH$_2$Cl$_2$ (10 mL) at −78 °C, and the resulting solution was stirred for 1 h. After this time, the reaction was quenched by pouring into a vigorously stirred mixture of EtOAc and potassium sodium tartrate solution. Once the two clear layers had formed, they were separated, and the aqueous phase was extracted three times with EtOAc. The combined organic extracts were washed with water and brine, dried (magnesium sulfate), and concentrated under reduced pressure. The crude lactol was dissolved in CH$_2$Cl$_2$ (10 mL) and cooled to −78 °C. Triethylsilane (0.55 mL, 3.42 mmol), followed by boron trifluoride etherate (0.29 mL, 2.28 mmol), were added, and the resulting solution was allowed to stir for 20 min. The reaction was quenched with saturated aqueous ammonium chloride. The layers were separated, and the aqueous phase was extracted three times with CH$_2$Cl$_2$. The combined organic extracts were washed with saturated aqueous NaHCO$_3$ and brine, dried (MgSO$_4$), and concentrated under reduced pressure. The crude product was purified by flash column chromatography eluting with hexanes:EtOAc (10:1) to yield 435 mg (77%) of the cyclic ether as a clear colorless oil.

Reference: Ahmed, A.; Hoegenauer, E. K.; Enev, V. S.; Hanbauer, M.; Kaehlig, H.; Öhler, E.; Mulzer, J. *J. Org. Chem.* **2003**, *68*, 3026–3042.

## 4.6.6 Shapiro Reaction

The Shapiro reaction is the conversion of ketone to an alkene via a tosylhydrazone. The tosylhydrazone is treated with two equivalents of strong base, typically *n*-BuLi or MeLi, to afford an alkenyllithium species which is quenched with an electrophile to form an alkene (see also chapter 5, 5.1.25 Shapiro reaction). Reviews: (a) Chamberlin, A. R.; Bloom, S. H. *Org. React.* **1990**, *39*, 1–83. (b) Shapiro, R.H. *Org. React.* **1976**, *23*, 405.

A suspension of the ketone (30.5 g, 0.10 mol), *p*-toluenesulfonylhydrazide (22.1 g, 0.119 mol), and *p*-tolunesulfonic acid monohydrate (4.92 g, 25. 9 mmol) in MeOH (611 mL) was heated at reflux under $N_2$ for 24 h. The resultant reaction slurry was cooled for 1 h in an ice-bath, filtered, and rinsed with cold MeOH (150 mL) to give 36.8 g (77%) of the hydrazone.

(*Caution: this transformation produces nitrogen gas. The reaction system should be vented. Do not run in a closed system! This reaction may also be run with diethyl ether.*) A suspension of the hydrazone (20.0 g, 43.2 mmol) in methyl *tert*-butylether (MTBE 400 mL) was treated with a solution of MeLi as a complex with LiBr (1.5 M in $Et_2O$, 86.3 mL, 0.13 mol) at room temperature for 1 h, cooled to 0 °C, and quenched with water (500 mL). The reaction was extracted with MTBE (1 L), the organic layer was dried ($MgSO_4$), and the solvent was removed in vacuo to give 12.0 g (99%) of the alkene as a white solid.

Reference: Faul, M. M.; Ratz, A. M.; Sullivan, K. A.; Trankle, W. G.; Winneroski, L. L. *J. Org. Chem.* **2001**, *66*, 5772–5782.

## 4.7 Reductive Dehalogenations

### 4.7.1 Tributyltin Hydride

Review: Neumann, W. P. *Synthesis* **1987**, 665–682.

#### 4.7.1.1 With 2,2'-Azobisisobutyronitrile

AIBN, a radical initiator

To a stirred solution of the iodolactone (296 mg, 0.74 mmol) in dry benzene (6 mL) was added *n*-tributyltin hydride (0.40 mL, 1.5 mmol) and a catalytic amount of AIBN. The mixture was heated at reflux for 30 min. It was then cooled to room temperature and concentrated to approximately 2 mL under reduced pressure. Flash chromatography (50 g of silica gel, 2:1 ether:petroleum ether then 4:1 $CH_2Cl_2$:ether of the crude product) provided 180 mg (89%) of the lactone as a colorless solid.

Reference: Britton, R. A.; Piers, E.; Patrick, B. O. *J. Org. Chem.* **2004**, *69*, 3068–3075.

#### 4.7.1.2 With Triethylborane

A solution of the dichlorolactone (1.5 g, 2.15 mmol) and freshly prepared tributyltin hydride (1.3 mL, 4.9 mmol) in benzene (20 mL) was heated at 78 °C for 15 min. in the presence of a catalytic amount of Et$_3$B (0.7 mL of a 1 M solution in THF, 0.7 mmol). After evaporation of the solvent in vacuo, the residue was dissolved in ether (25 mL). The ether solution was treated with a saturated aqueous solution of potassium fluoride (25 mL). The precipitated tri-*n*-butyltin fluoride was filtered and washed thoroughly with ether. The combined ether layers were dried over magnesium sulfate and evaporated in vacuo. The residue was chromatographed on silica gel using a gradient of eluents (hexanes:EtOAc, 20:1 to 4:1) giving 1.27 g (92%) of the lactone as a clear oil.

Reference: Marino, J. P.; Rubio, M. B.; Cao, G.; de Dios, A. *J. Am. Chem. Soc.* **2002**, *124*, 13398–13399.

### 4.7.2 Trialkylsilanes

Reviews: (a) Chatgilialoglu, C. *Organosilanes in Radical Chemistry Principles, Methods, and Applications*; Wiley&Sons, Ltd: West Sussex, U.K., 2004. (b) Chatgilialoglu, C. *Acc. Chem. Res.* **1992**, *25*, 188–194.

(TMS)$_3$SiH (850 μL, 2.76 mmol) and AIBN (30 mg, 0.18 mmol) were added to a solution of the bromoformate (980 mg, 1.84 mmol) in toluene (90 mL), and the resulting mixture was stirred at 80 °C for 4 h. The solution was allowed to reach ambient temperature before the solvent was evaporated. The residue was dissolved in methanol (100 mL). Aqueous saturated NaHCO$_3$ (about 12 mL) was added dropwise, and the reaction mixture was stirred for 2 h before it was diluted with water (25 mL). A standard extractive workup with MTBE followed by flash chromatography (hexanes:EtOAc, 8:1) of the crude product provided 705 mg (90%) of the alcohol as a colorless oil.

Reference: Lepage, O.; Kattnig, E.; Fürstner, A. *J. Am Chem. Soc.* **2004**, *126*, 15970–15971.

## 4.8 Carbon–Carbon Double and Triple Bond Reductions

### 4.8.1 Pd/C, H$_2$

The unsaturated [3.3.1] bicycle (360 mg, 0.78 mmol), 10% Pd/C (130 mg, 0.12 mmol), and EtOAc (8 mL) were combined, and the reaction vessel was evacuated and back-filled with hydrogen (1 atm). The reaction mixture was stirred under hydrogen for 30 min and then filtered over a plug of silica gel topped with Celite (EtOAc eluent) to afford 358 mg (99%) of the saturated bicycle as a colorless oil.

Reference: Garg, N. K.; Caspi, D. D.; Stoltz, B. M. *J. Am. Chem. Soc.* **2004**, *126*, 9552–9553.

### 4.8.2 Lindlar's Catalyst

Converts alkynes to *cis*-olefins

A suspension of Lindlar's catalyst [Pd (5% on $CaCO_3$), 213 mg, 0.10 mmol], quinolone (0.81 mL, 6.8 mmol), and the alkyne (587 mg, 2.19 mmol) in EtOAc (22 mL) was stirred for 30 min. The flask was charged with hydrogen (1 atm), and the reaction mixture was stirred until $^1$H NMR analysis indicated complete conversion (14 h). The reaction mixture was filtered through Celite and concentrated in vacuo. The crude product was purified by flash chromatography on silica gel (elution with $EtOAc/CH_2Cl_2$, 1:5 then EtOAc) to give 487 mg (83%) of the *cis*-olefin as a white solid.

Reference: Wang, Y.; Janjic, J.; Kozmin, S. A. *J. Am. Chem. Soc.* **2002**, *124*, 13670–13671.

### 4.8.3 Diimide

Diimide must be generated in situ. It is an effective reagent for the *syn*-reduction of alkenes and favors reduction of the more substituted olefin. Reviews: (a) Pasto, D. J. In *Comprehensive Organic Synthesis*; Trost, B. M.; Fleming, I., Eds.; Pergamon Press: Oxford, U. K., 1991; Vol. 8, Chapter 3.3: Reductions of C=C and C≡C by Noncatalytic Chemical Methods, pp. 472–478. (b) Pasto, D. J.; Taylor, R. T. *Org. React.* **1991**, *40*, 91–155.

Dipotassium azodicarboxylate (DAPA) (20 mg, 0.123 mmol) was added to a solution of the isomeric olefin (149 mg, 0.197 mmol) in methanol (10 mL). Glacial acetic acid (10 µL) was added, and the reaction mixture was heated to 40 °C open to the atmosphere. As the yellow color of the solution started to fade, more DAPA (20 mg) and acetic acid (10 µL) were added. The addition procedure was repeated several times over the course of 10 h, and the reaction progress was monitored by NMR. On completion of the reaction, the reaction mixture was cooled and then a saturated aqueous solution of NaHCO$_3$ (15 mL) was added. The product was then extracted into CH$_2$Cl$_2$ (3 × 25 mL), and the combined organic layers were dried over Na$_2$SO$_4$. After evaporation of the solvent, the clear residue was purified by chromatography (2:1, hexane:EtOAc) to give 130 mg (87%) of the reduced product as a clear oil.

Reference: Centrone, C. A.; Lowary, T. L. *J. Org. Chem.* **2002**, *67*, 8862–8870.

### 4.8.4 Sodium Bis(2-methoxyethoxy(aluminum hydride))

Sodium bis(2-methoxyethoxy(aluminum hydride)) (Redal-H) is an excellent reagent for reducing propargylic alcohols to allylic alcohols. Reviews: (a) Seyden-Penne, J. *Reductions by the Alumino- and Borohydrides in organic Synthesis*; Wiley-VCH: New York, 1997, 2$^{nd}$ edition. (b) Málek, J. *Org. React.* **1985**, *34*, 1–317.

To a solution of the propargylic alcohol (7.7 g, 40 mmol) in ether (150 mL) was added Redal–H (18 mL of a 65% solution in toluene, 60 mmol) at 0 °C, and the mixture was stirred for 2 h at 0 °C. The mixture was then stirred at room temperature for 3 h, and the reaction was quenched with saturated potassium sodium tartrate solution at 0 °C. After separation of the mixture, the aqueous washes were extracted with EtOAc (four times). The combined organic layers were washed with brine, dried (magnesium sulfate), and concentrated in vacuo. The crude product was purified by silica gel column chromatography (EtOAc/hexanes 20%, 25%) to give 7.3 g (94%) of the *trans*-allylic alcohol as a yellow oil.

Reference: Nakamura, R; Tanino, K.; Miyashita, M. *Org. Lett.* **2003**, *5*, 3579–3582.

### 4.8.5 Lithium/Ammonia (Birch Reduction)

The Birch reduction is a dissolved metal reduction where a 1-electron transfer reduction of an aromatic ring affords a carbanion. The carbanion may be quenched with a proton source or an electrophile. Reviews: (a) Rabideau, P.W.; Marcinow, Z. *Org. React.* **1992**, *42*, 1–334. (b) Mander, L. N. In *Comprehensive Organic Synthesis*; Trost, B. M.; Fleming, I., Eds.; Pergamon Press: Oxford, U. K., 1991; Vol. 8, Chapter 3.4: Partial Reduction of Aromatic Rings by Dissolving Metals and by Other Methods, pp. 490–514. (c) Rabideau, P. W. *Tetrahedron* **1989**, *45*, 1579–1603.

To a solution of the isoquinolinone (1.156 g, 9.90 mmol) and *tert*-butyl alcohol (0.88 mL, 11.9 mmol) in THF (30 mL) at −78 °C was added liquid ammonia (about 280 mL). Lithium was added in small pieces until the blue coloration persisted, after which the solution was stirred at −78 °C for 30 min. The blue coloration was dissipated with piperlyne, 4-methoxybenzyl chloride (4.83 g, 31.00 mmol) in THF (5 mL) was introduced by syringe, and the mixture was stirred for an additional 150 min at −78 °C. Solid ammonium chloride was added and then the ammonia was allowed to evaporate. The pale yellow residue was partitioned between $CH_2Cl_2$ (30 mL) and water (40 mL). The layers were separated, and the aqueous layer was extracted with $CH_2Cl_2$ (2 × 30 mL). The combined organic layers were washed with 10% sodium thiosulfate solution (20 mL), dried over magnesium sulfate, and concentrated. Flash chromatography (EtOAc:hexane, 2:1) on silica gave 2.21 g (75%) of the tetrahydroisoquinolinone.

Reference: Schultz, A. G.; Guzi, T. J.; Larsson, E.; Rahm, R.; Thakker, K. Bidlack, J. M. *J. Org. Chem.* **1998**, *63*, 7795–7804.

## 4.9 Heteroatom–Heteroatom Reductions

### 4.9.1 Reduction of a Nitro Group to an Amine Using Ammonium Formate

A mild method for converting a nitro group to an amino functionality is by using the hydrogen transfer reagent anmonium formate in the presence of Pd/C. Reviews: (a) Kabalka, G. W.; Varma, R. S. In *Comprehensive Organic Synthesis*; Trost, B. M.; Fleming, I., Eds.; Pergamon Press: Oxford, U. K., 1991; Vol. 8, Chapter 2.1: Reduction of Nitro and Nitroso Compounds, pp. 363–379. (b) Ram, S.; Ehrenkaufer, R.E. *Synthesis* **1988**, 91–94.

## REDUCTIONS

$O_2N\text{-CH}_2\text{-CH(OH)-CH}_2\text{-CH(NHBOC)-CO}_2t\text{-Bu}$ →[Pd/C, NH$_4^+$COO$^-$ / MeOH, −10 °C] $H_2N\text{-CH}_2\text{-CH(OH)-CH}_2\text{-CH(NHBOC)-CO}_2t\text{-Bu}$

100%

The nitroalcohol (5 g, 14.9 mmol) is dissolved in methanol (50 mL). The reaction mixture is cooled to −10 °C, and palladium on carbon (10%, purissimum, Fluka, 2.5 g) and dry ammonium formate (9.43 g, 150 mmol) are added while maintaining the reaction temperature at −10 °C. After stirring the reaction for 2 h, the catalyst is filtered off. The solvent is removed and EtOAc and saturated aqueous NaHCO$_3$ solution are added (pH ≥ 7). The phases are separated and after two additional washings with EtOAc, the combined organic phases are washed with brine, dried over magnesium sulfate, and the solvent is removed in vacuo. The amino alcohol is obtained as a colorless yellow oil in a crude yield of 4.55 g (100%).

Reference: Rudolph, J.; Hanning, F.; Theis, H.; Wischnat, R. *Org. Lett.* **2001**, *3*, 3153–3155.

### 4.9.2 Sodium Borohydride/Nickel Dichloride

[Decalin structure with NTs, HO, N$_3$, isopropyl substituents] →[NaBH$_4$, NiCl$_2$ / MeOH, rt] [Decalin structure with NTs, HO, H$_2$N, isopropyl substituents]

98%

To a solution of the azide (617 mg, 1.43 mmol) in methanol/THF (30 mL, 3:1) at 0 °C was added nickel dichloride hexahydrate (542 mg, 2.28 mmol) followed by sodium borohydride (248 mg, 6.57 mmol) over 10 min. After 30 min, the black mixture was allowed to warm to 25 °C, diluted with EtOAc (40 mL), and filtered through Celite. The solution was further diluted with EtOAc (50 mL) and washed with brine (2 × 25 mL) and 0.01 M EDTA solution (1 × 25 mL, pH 7.5, K-phosphate buffer). After solvent removal in vacuo, the resulting oil was purified by silica gel chromatography (100% EtOAc then EtOAc:methanol, 12:1) to give 569 mg (98%) of the amine.

Reference: White, R. D.; Wood, J. L. *Org. Lett.* **2001**, *3*, 1825–1827.

### 4.9.3 Staudinger Reaction

The Staudinger reaction is a mild method for the conversion of azides to amines by treating the azide with triphenylphosphine in the presence of water. The water is necessary, because the intermediate iminophosphrane is hydrolyzed to form the amine and triphenylphosphine oxide. Reviews: (a) Gololobov, Y. G.; Kasukhin, L. F. *Tetrahedron* **1992**, *48*, 1353–1406. (b) Gololobov, Y. G.; Zhmurova, I. N.; Kasukhin, L. F. *Tetrahedron* **1981**, *37*, 437–472.

BnO~~CH(OMOM)CH(N₃)CH₃ →[PPh₃, THF, rt, 81%] BnO~~CH(OMOM)CH(NH₂)CH₃

To a solution of the azide (0.45 g, 1.4 mmol) in THF/water (10:1.5 mL) was added triphenylphosphine (0.41g, 1.6 mmol) at room temperature. The reaction mixture was then stirred for 24 h. The mixture was concentrated, and the residue was purified by preparative TLC eluted with $CHCl_3$:MeOH (18:1) to afford 0.33 g (81%) of the amine as a colorless oil.

Reference: Makabe, H.; Kong, L. K.; Hirota, M. *Org. Lett.* **2003**, *5*, 27–29.

# 5

# Carbon–Carbon Bond Formation

## 5.1 Carbon–Carbon Forming Reactions (Single Bonds)

### 5.1.1 Aldol Reactions

Reviews: (a) Vicarion, J. L.; Badia, D.; Carillo, L.; Reyes, E.; Etxebarria, J. *Curr. Org. Chem.* **2005,** *9*, 219–235. (b) Mahrwald, R. Ed. In *Modern Aldol Reactions*; Wiley-VCH: Weinheim, 2004; Vol. 1., pp. 1–335 (c) Mahrwald, R. Ed. In *Modern Aldol Reactions*; Wiley-VCH: Weinheim, 2004; Vol. 2., pp. 1–345.(d) Machajewski, T. D.; Wong, C.-H. *Angew. Chem. Int. Ed.* **2000,** *39*, 1352–1375. (e) Carriera, E. M. In Modern Carbonyl Chemistry; Otera, J.; Wiley-VCH: Weinheim, 2000; Chapter 8: Aldol Reaction: Methodology and Stereochemistry, 227–248. (f) Paterson, I.; Cowden, C. J.; Wallace, D. J. In Modern Carbonyl Chemistry; Otera, J.; Wiley-VCH: Weinheim, 2000; Chapter 9: Stereoselective Aldol Reactions in the Synthesis of Polyketide Natural Products, pp. 249–298. (g) Franklin, A. S.; Paterson, I. Contemp. Org. Synth. **1994,** *1* 317–338. (h) Heathcock, C. H. In *Asymmetric Synthesis*; Morrison, J. D., Ed.; Academic Press: Orlando, Fl.; 1984; Vol. 3., Chapter 2: The Aldol Addition Reaction, pp. 111–212. (i) Mukaiyama, T. *Org. React.* **1982,** *28*, 203–331.

#### 5.1.1.1 With Boron Enolates

Since the early 1980s, aldol condensations involving boron enolates have gain great importance in asymmetric synthesis, particularly the synthesis of natural products with adjacent stereogenic centers bearing hydroxyl and methyl groups. (*Z*)-Boron enolates tend to give a high diastereoslectivity preference for the *syn*-stereochemistry while (*E*)-boron enolates favor the *anti*-stereochemistry. Because the B–O and

B–C bonds are shorter than other metals with oxygen and carbon, the six membered Zimmerman–Traxler transition state in the aldol condensation tends to be more compact which accentuates steric interactions, thus leading to higher diastereoselectivity. When this feature is coupled with a boron enolate bearing a chiral auxillary, high enantioselectivity is achieved. Boron enolates are generated from a ketone and boron triflate in the presence of an organic base such as triethylamine. Reviews: (a) Abiko, A. *Acc. Chem. Res.* **2004**, *37*, 387–395. (b) Cowden, C. J. *Org. React.* **1997**, *51*, 1–200.

### 5.1.1.1.1 Evans Aldol (With Two Oxazolidinone Chiral Auxillaries)

A 250-mL round-bottomed flask was charged with the oxazolidinone (3.24, 9.37 mmol) in $CH_2Cl_2$ (20 mL). The solution was cooled in an ice bath and then dibutylboron triflate (2.83 mL, 11.3 mmol) was added dropwise, followed by triethylamine (1.70 mL, 12.2 ml). The resulting yellow solution was stirred at 0 °C for 10 min. The ice bath was replaced with a dry ice–acetone bath (−78 °C), and freshly distilled hexanal (1.35 mL, 11.3 mmol) was added by syringe over a 5-min period. The reaction mixture was stirred at −78 °C for 1 h, warmed to 0 °C for 2 h, and quenched by addition of pH 7 buffer (15 mL) and methanol (15 mL). A mixture of methanol:30% aqueous $H_2O_2$ solution (2:1, 10 mL) was then added dropwise, and the biphasic mixture was stirred vigorously at room temperature for 1 h. The mixture was diluted with water (100 mL) and the layers were separated. The aqueous layer was washed with $CH_2Cl_2$ (2×), and the combined organic layers were washed with saturated aqueous sodium bicarbonate, dried over magnesium sulfate, and concentrated. The crude product was purified by flash chromatography ($SiO_2$) eluting with hexanes:EtOAc (85:15) to afford 3.62 g (97%) of the alcohol as a pale yellow oil.

Reference: Durham, T. B.; Blanchard, N.; Savall, B. M.; Powell, N. A.; Roush, W. R. *J. Am. Chem. Soc.* **2004**, *126*, 9307–9317.

Di-*n*-butylboron-triflate (4.71 g, 18.9 mmol) was added dropwise with stirring to a solution of the imide (4.50 g, 19.3 mmol) in $CH_2Cl_2$ (35 mL) at −78 °C. The cooling bath was removed, and the resulting mixture was stirred at room temperature until the solution became homogeneous. The mixture was re-cooled to −78 °C, and triethylamine (3.14 mL, 22.5 mL) was added dropwise. After 0.5 h at −78 °C, the reaction was warmed to 0 °C, maintained at this temperature for 1 h, and re-cooled to −78 °C. A solution of the enaldehyde (2.34 g, 6.60 mmol) in tetrahydrofuran (THF 30 mL) was added via a cannula, and the mixture was stirred for 0.5 h at −78 °C and then at 0 °C for 1 h. Phosphate buffer (30 mL of 0.25 M, pH = 7) and then 30% hydrogen peroxide in methanol (60 mL, 1:1 v/v) were added. The resulting cloudy mixture was diluted with a sufficient volume of methanol to produce a nearly homogeneous solution and subsequently stirred at 0 °C for 1 h. The majority of the methanol (about 75 mL) was removed under reduced pressure (*caution*: the water bath temperature was < 30 °C), and the resulting solution was extracted with $CH_2Cl_2$ (3 × 50 mL). The combined organic extracts were washed with saturated sodium bicarbonate (1 × 50 mL), dried ($MgSO_4$), and concentrated under reduced pressure to afford a pale yellow oil that was purified by flash chromatography by gradient elution, first with hexanes:ether (7:3) to elute excess oxazolidinone and then with hexanes:ether (1:1) to give 3.50 g (91%) of the aldol adduct as a white foam.

Reference: Martin, S. F.; Dodge, J. A.; Burgess, L. E.; Limberakis, C.; Hartmann, M. *Tetrahedron* **1996**, *52*, 3229–3246.

### 5.1.1.1.2 Masamune Aldol

The Masamune aldol condensation, in common with the Evans aldol condensation, involves a boron enolate of an ester containing a norephedrine derived chiral auxillary; however, unlike the latter, the Masamune aldol delivers a 3-hydroxy-2-methyl carbonyl moiety with the *anti*-stereochemistry. Crucial to the success of this reaction is the use of dicyclohexylboron triflate to generate the boron enolate. *Note: in the Evans' aldol condensation, dibutylboron triflate is utilized.*

To a stirred solution of the ester (734 mg, 1.53 mmol) and triethylamine (0.52 mL, 3.67 mmol) in $CH_2Cl_2$ (8 mL) at −78 °C was added dropwise a solution of dicyclohexylboron triflate (1.0 M in hexane, 3.4 mL, 3.4 mmol) over 20 min. The resulting mixture was stirred at this temperature for 2 h before a solution of the aldehyde (442 mg, 1.84 mmol) in $CH_2Cl_2$ (15 mL) was introduced. Stirring was continued for 1 h at −78 °C before the mixture was allowed to reach ambient temperature over a period of 2 h. The reaction was quenched with a pH 7 buffer solution (6 mL), the mixture was diluted with methanol (21 mL), and 30% hydrogen peroxide (3.1 mL) was added carefully. The mixture was then vigorously stirred overnight before it was concentrated. The residue was partitioned between water and $CH_2Cl_2$, and the combined organic phases were dried over sodium sulfate and evaporated. The residue was purified by flash chromatography eluting with a gradient of hexane:EtOAc (30:1 to 9:1) to give 862 mg (71%) over the alcohol as a colorless syrup.

References: Fürstner, A.; Caro-Ruiz, J.; Prinz, H.; Waldmann, H. *J. Org. Chem.* **2004,** *69,* 459–467. Abiko, A. Synthesis of the ester: *Org. Syn.* **2002,** *79,* 109–113.

### 5.1.1.2 Mukaiyama Aldol

The enol silane can be prepared from aldehydes, ketones, esters, and thioesters. The asymmetric Mukaiyama aldol reaction has also been developed using chiral substrates and Lewis acids.

#### 5.1.1.2.1 Standard

$$\text{PhC(OTMS)=CH}_2 + \text{acetone} \xrightarrow[\text{CH}_2\text{Cl}_2,\ 0\ °\text{C}]{\text{TiCl}_4} \text{PhCOCH}_2\text{C(CH}_3)_2\text{OH}$$

70–74%

*Synthesis of the silylenol ether*

To a four-necked flask, equipped with a mechanical stirrer, a reflux condenser with a nitrogen inlet, a thermometer, and a pressure-equalizing dropping funnel, was added 36 g (0.30 mol) of acetone and 41.4 g (0.41 mol) of triethylamine under a nitrogen atmosphere. To this mixture, stirred at room temperature under nitrogen, was added via the dropping funnel 43.2 g (0.40 mol) of chlorotrimethylsilane over 10 min. The flask was then immersed in a water bath and the contents were warmed to 35°C. The water bath was removed and the dropping funnel was charged with a solution of 60 g (0.40 mol) of sodium iodide in 350 mL of acetonitrile. This solution was added to the stirred mixture in the flask at such a rate that the temperature of the reaction was maintained at 34–40 °C without external heating or cooling. The addition required approximately 1 h. When the addition was complete, the reaction mixture was stirred for a further 2 h at room temperature. The contents of the flask were then poured into ice-cold water, and the aqueous mixture was extracted three times with pentane. After extraction, the organic layer was dried over potassium carbonate and then concentrated with a rotary evaporator under reduced pressure. The crude product was a mixture of 97% of the desired silyl enol ether and 3% of acetophenone, as shown by gas chromatography. The crude product was distilled in a Claisen flask at a pressure of about 40 mm. After a small forerun (about 3 g), 52.3 g (91%) of silyl enol ether, bp 124–125.5°C, was obtained. The purity of this material was approximately 98%, as judged by gas chromatography and $^1$H NMR spectroscopy.

*Aldol*

A 500-mL three-necked flask was fitted with a stirring bar, a rubber stopper, a 100-mL pressure-equalizing dropping funnel, and a three-way stopcock that is equipped with a balloon of argon gas. The flask was charged with dry methylene chloride (140 mL) and cooled in an ice bath. Titanium tetrachloride (11.0 mL) was added by a syringe with stirring by a magnetic stirrer, and a solution of 6.5 g of acetone in methylene chloride (30 mL) was added dropwise over a 5-min period. On completion of this addition a solution of 19.2 g of acetophenone trimethylsilyl enol ether in methylene chloride (15 mL) was added dropwise over a 10-min period, and the mixture was stirred for 15 min.

The reaction mixture is poured into ice water (200 mL) with vigorous stirring and the organic layer was separated. The aqueous layer was extracted with two 30-mL portions of methylene chloride. The combined methylene chloride extracts were washed with two 60-mL portions of a 1:1 mixture of saturated aqueous sodium bicarbonate and water, and then with brine. The methylene chloride solution was dried over sodium sulfate and the methylene chloride was removed using a rotary evaporator.

The residue was dissolved in 30 mL of benzene, and the solution was transferred to a chromatographic column (50-mm diameter) consisting of 600 mL of silica gel. The product was eluted sequentially with (a) 1 L of 4:1 (v/v) hexane:EtOAc and (b) 1.5 L of 2:1 (v/v) hexane:EtOAc (flash chromatography). The initial 900 mL of the eluent was discarded. Concentration of the later fractions (about 1.3 L) under reduced pressure yielded the pure product as an oil. The total yield was 12.2–12.8 g (70–74%).

Reference: Mukaiyama, T.; Narasaka, K. *Org. Syn.* **1993**, *Coll. Vol. 8*, 323.

### 5.1.1.2.2 Asymmetric Mukaiyama

To the aldehyde (0.100g, 0.494 mmol) in $CH_2Cl_2$ (5 mL) was added (1-*tert*-butylsulfanyl-vinyloxy)-trimethylsilane (131 mg, 0.641 mmol), and the flask was immersed in a −78 °C cooling bath. To the mixture was added a solution of dimethylaluminum chloride (1.0 M in hexanes, 0.74 mL, 0.74 mmol) dropwise, and the reaction was stirred for 1 h at −78 °C. The reaction was quenched by the addition of 10% w/v citric acid in MeOH and slowly raised to ambient temperature; stirring was continued for 1 h. The reaction was diluted with water, and the aqueous layer was extracted with ether (3 ×). The combined organic layers were dried over sodium sulfate, filtered, and the crude product mixture was concentrated in vacuo. The remaining volatiles were removed under high vacuum. Purification via flash chromatography ($SiO_2$, 10:1 hexanes/EtOAc) afforded 143 mg (86%) of the β-hydroxythioester as a clear viscous oil.

Reference: Stevens, B. D.; Bungard, C. J.; Nelson, S. G. *J. Org. Chem.* **2006**, *71*, 6397–6402.

### 5.1.1.3 Crimmins' Asymmetric Aldol (Thione Chiral Auxillary)

TiCl$_4$ (0.44 mL, 3.99 mmol) was added dropwise to a solution of the thione (1.061g, 3.32 mmol) in dry CH$_2$Cl$_2$ (16 mL) and stirred at 0 °C under N$_2$. After 10 min, dry $N,N,N'N'$-tetramethylethylenediamine (TMEDA 1.65 mL, 8.3 mmol) was added dropwise. The dark solution was then stirred at 0 °C for 0.5 h before the aldehydes (1.540 g, 6.64 mmol) were introduced dropwise. The reaction mixture was stirred at the same temperature for 2 h. Aqueous NH$_4$Cl (30 mL) was introduced, followed by diethyl ether (300 mL). The phases were separated, and the organic phase was washed with aqueous NH$_4$Cl (2 × 50 mL) and water, dried (Na$_2$SO$_4$), and filtered. The solvents were removed, and the resulting oily residue was chromatographed on silica gel eluting with n-hexane/EtOAc (9:1 to 3:1) to give 1.422 g (78%) of the *syn*-aldol adduct as a pale yellow sticky oil.

Reference: Wu, Y.; Sun, Y.-P. *J. Org. Chem.* **2006**, *71*, 5748–5751.

### 5.1.2 Asymmetric Deprotonation

An oven-dried, 2-L, three-necked flask, equipped with a magnetic stir bar and a thermocouple, was charged with (−)-sparteine (30.2 mL, 131 mmol), *N*-Boc-pyrrolidine (15.0 g, 87.6 mmol), and anhydrous ether (900 mL). The solution was cooled to −70 °C (dry ice/acetone bath). To this solution was added *sec*-butyllithium (96 mL, 1.16 M in cyclohexane, 111 mmol) dropwise over a period of 35 min. The reaction was then stirred at −70°C for 5.5 h.

After this interval, a solution of benzophenone (25.5 g, 140 mmol) in anhydrous ether (200 mL) was added dropwise over a period of 1.25 h. The dark green to greenish-yellow suspension was maintained at −70 °C for 2.0 h, and the reaction was then quenched by dropwise addition of glacial acetic acid (8.5 mL, 150 mmol) over a period of 15 min. The resulting lemon-yellow suspension was allowed to warm slowly to room temperature over a period of 12 h, during which time the mixture becomes cream colored.

After the solution was warmed to 25 °C, 5% phosphoric acid ($H_3PO_4$) (150 mL) was added to the reaction mixture, and the resulting biphasic mixture was stirred for 20 min. The layers were partitioned, and the organic phase was washed with additional 5% $H_3PO_4$ (3 × 150 mL). Combined aqueous phases were extracted with ether (3 × 200 mL). The original organic phase and the ethereal extracts were combined, washed with brine (200 mL), dried over magnesium sulfate ($MgSO_4$), filtered, and the solvents were removed under reduced pressure to afford crude product as an off-white solid. The crude (R)-(+)-2-(diphenylhydroxymethyl)-N-(tert-butoxycarbonyl)pyrrolidine was purified by recrystallization from a mixture of hexanes-EtOAc (~ 675 mL, 20:1, v/v) affording in two crops 20.9–22.0 g (73–74%) of analytically pure product as a white solid having greater than 99.5% *ee*.

Sparteine was recovered by making the aqueous phases basic with aqueous 20% sodium hydroxide (160 mL). The aqueous phase was extracted with $Et_2O$ (4 × 150 mL), and the combined organic phases were dried over potassium carbonate ($K_2CO_3$), filtered, and the solvents removed under reduced pressure to afford 30.3 g (98%) of crude recovered sparteine as a pale yellow oil. Fractional distillation of the residual oil from calcium hydride ($CaH_2$) afforded 27.0 g of sparteine (88%) suitable for reuse.

Reference: Nikolic, N. A.; Beak, P. *Org. Syn.* **1998**, *Coll. Vol. 9*, 391–397.

### 5.1.3 Baylis–Hillman Reaction

The Baylis–Hillman reaction has become a very powerful carbon–carbon bond forming reaction in the past 20 years. A typical reaction involves an activated olefin (i.e., an acrylate) and an aldehyde in the presence of a tertiary amine such as diazobicyclo-[2.2.2]octane (DABCO) to form an α-methylhydroxyacrylate. A host of activated olefins have been utilized including acrylates, acroleins, α,β-unsaturated ketones, vinylsulfones, vinylphosphonates, vinyl nitriles, *etc*. The Baylis–Hillman has been successfully applied inter- and intramolecularly. In addition, there are numerous examples of asymmetric Baylis–Hillman reactions. Reviews: (a) Ciganek, E. *Org. React.* **1997**, *51*, 201–478. (b) Basavaiah, D.; Rao, P. D.; Hyma, R. S. *Tetrahedron* **1996**, *52*, 8001–8062. (c) Fort, Y.; Berthe, M. C.; Caubere, P. *Tetrahedron* **1992**, *48*, 6371–6384.

To a stirred solution of the aldehyde (1.0 g, 2.92 mmol) in THF (6 mL) were added DABCO (0.15 g, 1.32 mmol) and ethyl acrylate (0.8 mL, 7.28 mmol). The reaction mixture was stirred at room temperature for 24 h. The solvent was removed to give a residue which was purified by chromatography (silica gel) eluting with hexane:EtOAc (8.5:1.5) to afford 1.62 g (97%) of the ethyl acrylate as a syrup.

Reference: Krishna, P. R.; Kannan, V.; Ilangovan, A.; Sharma, G. V. M. *Tetrahedron: Asymmetry* **2001**, *12*, 829–837.

### 5.1.4 Benzoin Condensation

In the traditional benzoin condensation, aromatic and heteroaryl aldehydes are converted to α-hydroxy ketones in the presence of a catalytic amount of cyanide ion. Since the discovery of this reaction in the early 20th century, improvements have made, including the use of quaternary thiazolium salts as the catalyst. This improvement has broadened the scope of this reaction to include aliphatic aldehydes to deliver acyloins. Review: Hassner, A.; Lokanatha Rai, K. M. In *Comprehensive Organic Synthesis*, Trost, B. M.; Fleming, I. Eds.; Pergamon Press: Oxford, 1991; Vol. 1, Chapter 2.4: The Benzoin and Related Acyl Anion Equivalent Reactions, pp. 541–577.

A 500 ml, three-necked, round-bottomed flask was equipped with a mechanical stirrer, a short gas inlet tube, and an efficient reflux condenser fitted with a potassium hydroxide drying tube. The flask was charged with 3-benzyl-5-(2-hydroxyethyl)-4-methyl1,3-thiazolium chloride (13.4 g, 0.05 mol), 72.1 g (1.0 mol) of butyraldehyde, triethylamine (30.3 g, 0.3 mol), and absolute ethanol (300 mL). A slow stream of nitrogen was begun, and the mixture was stirred and heated in an oil bath at 80 °C. After 1.5 h the reaction mixture was cooled to room temperature and concentrated by rotary evaporation. The residual yellow liquid was poured into $CH_2Cl_2$ (150 mL) which was then used to extract the aqueous mixture. The aqueous layer was extracted with another portion of $CH_2Cl_2$ (150 mL). The combined organic layers were washed with saturated sodium bicarbonate (300 mL) and water (300 mL). The organic layer was then concentrated under reduced pressure. The crude product was purified via distillation through a 20-cm Vigreux column to afford 51–54 g (71–74%) of the product as a light-yellow liquid.

Reference: Stetter, H.; Kuhlmann, H. *Org. Syn.* **1990**, *Coll. Vol. VII*, 95–99.

### 5.1.5 Brown Asymmetric Crotylation

The Brown asymmetric crotylation is a highly regioselective and stereospecific reaction. Many organoboranes are now commercial available. Reviews: (a) Denmark, S. E.; Almstead, N. G. In *Modern Carbonyl Chemistry*; Otera, J, Ed.; Wiley-VCH: Weinheim, 2000; Chapter 10: Allylation of Carbonyls: Methodology and Stereochemistry, pp. 299–402. (b) Chemler, S. R.; Roush, W. R. In *Modern Carbonyl Chemistry*; Otera, J.; Wiley-VCH: Weinheim, 2000; Chapter 11: Recent Applications of the Allylation Reaction to the Synthesis of Natural Products, pp. 403–490.

TBSO–CH$_2$CH$_2$CHO

1. (*Z*)-butene, *n*-BuLi, KO*t*-Bu, −78 to −45 °C
2. (+)-Ipc$_2$BOMe, BF$_3$•OEt$_2$, −78 °C to rt
3. NaOH, rt

→ TBSO–CH$_2$CH$_2$CH(OH)CH(CH$_3$)CH=CH$_2$

89%, *ds* > 99:1, *ee* = 96%

To a solution of potassium *tert*-butoxide (2.84 g, 24.0 mmol) in THF (40 mL) at −78 °C was cannulated a solution of *cis*-butene (15 mL, 167 mmol). *n*-BuLi (1.6 M in hexanes, 15.5 mL, 24.8 mmol) was then added dropwise, and the yellow mixture was stirred at −78 °C for 5 min and at −45 °C for 20 min. The resulting orange solution was cooled to −78 °C, and a solution of (+)-*B*-diisopinocamphenylmethoxyborane (9.05 g, 28.6 mmol) in diethyl ether (20 mL) was added dropwise over approximately 20 min. The resulting white solution was stirred at −78 °C for 45 min. Boron trifluoride etherate (4.4 mL, 34.2 mmol) was added dropwise followed after 5 min by addition of a solution of 3-(*tert*-butyldimethyl-silyloxy)propionaldehyde (3.76 g, 20.0 mmol) in THF (12 mL). The resulting solution was stirred at room temperature for 15 h. The reaction mixture was neutralized with dilute HCl and washed with diethyl ether. The combined organic layers were washed with brine, dried (sodium sulfate), filtered, and concentrated under reduced pressure. The resulting crude product was purified via flash chromatography on silica gel eluting with hexane:EtOAc (9:1) to deliver 4.34 g (89%, 95% *ee*) of the homoallylic alcohol as a colorless oil.

Reference: Wang, P.; Kim, Y.-J.; Navarro-Villalobos, M.; Rohde, B. D.; Gin, D. Y. *J. Am. Chem. Soc.* **2005**, *127*, 3256–3257.

### 5.1.6 Claisen Condensation

1. LDA, THF, −45 to −50 °C
2. Cbz-HN-CH(CH$_2$Ph)-CO$_2$Me, −50 °C

97%

A solution of *tert*-butyl acetate (4.0 equivalents) in THF was added dropwise to a mixture of dry THF and lithium diisopropylamide (LDA 3.5 equivalents in THF) with stirring at − 45 °C. After stirring at −45 °C for 60 min, a solution of the methyl ester was added while maintaining the internal temperature below −50 °C. The resulting mixture was stirred at −50 °C for 60 min and then poured into 1 M HCl and extracted with toluene. The organic layer was then washed with 5% NaHCO$_3$, dried (MgSO$_4$), filtered, and concentrated under reduced pressure to afford a crude product. The crude product was purified by column chromatography to give 97% yield of the β-keto ester.

Reference: Honda, Y.; Katayama, S.; Kojima, M.; Suzuki, T.; Izawa, K. *Org. Lett.* **2002**, *4*, 447–449.

### 5.1.7 Cuprates

The stability of organocuprates depends on the temperature. For instance, $Ph_2CuLi$ is stable at room temperature while $Me_2CuLi$ is better used at 0 °C. For other unstable organocuprates, the reactions are preferably carried out at −30 °C or lower if the rates of the reactions allow. Reviews: (a) Krause, N.; Gerold, A. *Angew. Chem. Int. Ed.* **1997**, *36*, 186–204. (b) Lipshutz, B. H.; Sengupta, S. *Org. React.* **1992**, *41*, 135–631. (c) Lipshutz, B. H. *Synthesis* **1987,** 325–341. (d) Normant, J. F. *Synthesis* **1972**, 63–80.

#### 5.1.7.1 Lower Order Cuprates

To a −5 °C slurry of copper(I) iodide (9.83 g, 50.56 mmol) in ether (250 mL) was added dropwise methyl lithium (1.4 M in ether, 72.2 mL, 101.12 mmol). The resulting light-tan solution was stirred for 60 min at −5 °C before a solution of the enone (4.5 g, 25.28 mmol) in ether (100 mL) was added over 10 min via an addition funnel. The bright-yellow reaction mixture was stirred for an additional 10 min at −5 °C until all of the starting material appeared to be consumed by thin layer chromatography (TLC). After quenching with saturated $NH_4Cl/NH_4OH$ solution (250 mL), the layers were separated, and the aqueous layer was extracted with ether (2 × 250 mL). The combined organic layers were dried over magnesium sulfate and concentrated to a yellow oil. The crude product was purified by passing it through a silica gel plug eluting with hexanes:EtOAc (9:1) to afford 4.76 g (97%) of the ketone as a pale yellow oil.

Reference: Spessard, S. J.; Stoltz, B. M. *Org. Lett.* **2002**, *4*, 1943–1946.

#### 5.1.7.2 High Order Cuprates (Cyanocuprates)

To a stirred solution of *t*-BuLi (1.3 M in pentane, 48 mL, 62.4 mmol) in diethyl ether (63 mL) at −78 °C was added a solution of the vinyl iodide (10.27 g, 36.4 mmol) in diethyl ether (75 mL) via a syringe pump over 20 min. After 20 min, the slurry was rapidly transferred to a precooled solution of copper cyanide (1.58 g, 17.7 mmol) in THF (122 mL) at −78 °C. After 1 h at −78 °C and 5 min at −40 °C, the solution was re-cooled to −78 °C, and a precooled solution of the oxazolidinone (3.40 g, 14.7 mmol) in THF (86 mL) was added via a cannula. An additional amount of THF (25 mL) was added to rinse the flask. After 30 min, the solution was warmed to 0 °C, and after a further 10 min the reaction was quenched with saturated aqueous ammonium chloride (300 mL) and extracted with ether (3 × 150 mL). The organic extracts were dried (magnesium sulfate) and concentrated in vacuo. The crude product was purified by chromatography (silica gel) eluting with ether: petroleum ether (15:85 to 50:50) to give 5.05 g (89%) of the oxazolidinone as a colorless oil.

Reference: White, J. D.; Carter, R. G.; Sundermann, K. F. *J. Org. Chem.* **1999**, *64*, 684–685.

### 5.1.8 Dieckmann Condensation

The Dieckmann condensation is an intramolecular variant of the Claisen condensation where a diester is converted to a β-ketoester. Typically, an alkoxide is used as the base to form the enolate which attacks the remaining ester to form the carbocycle. Five- and six-membered rings are formed readily with this method. Reviews: (a) Davis, B. R.; Garrett, P. J. In *Comprehensive Organic Synthesis,* Trost, B. M.; Fleming, I. Eds.; Pergamon Press: Oxford, 1991; Vol. 2, Chapter 3.6: Acylation of Esters, Ketones, and Nitriles, pp. 806–829. (b) Schaefer, J. P.; Bloomfield, J. *J. Org. React.* **1967**, *15*, 1–203.

Potassium *tert*-butoxide (4.3 g, 38.34 mmol) was added in 1 portion to a solution of the diester (8 g, 25.56 mmol) in toluene (80 mL) at 0 °C. After being stirred at the same temperature for 30 min, the reaction mixture was kept at room temperature overnight. Water (100 mL) was added, the phases were separated, and the aqueous phase was extracted with EtOAc (3 × 100 mL). The combined organic extracts were dried and evaporated. The residue was purified by flash chromatography eluting with EtOAc:light petroleum (1:9) to give 5.6 g (78%) of the keto-ester as an orange oil.

Reference: De Risi, C.; Pollini, G. P.; Trapella, C.; Peretto, I.; Ronzoni, S.; Giardina, G. A. M. *Bioorg. Med. Chem.* **2001**, *9*, 1871–1877.

## 5.1.9 Enolate Alkylations

### 5.1.9.1 Asymmetric Alkylation With Chiral Oxazolidinones

#### 5.1.9.1.1 Norephedrine-Derived Oxazolidinone

Sodium bis-trimethylsilylamide (1 M in THF, 6.4 mL, 6.4 mmol) was added dropwise to the imide (1.4 g, 5.7 mmol) in THF (15 mL) at −78 °C. The mixture was stirred at −78 °C for 2 h, allyl bromide (2.5 mL, 29 mmol) was then added via a syringe, and the mixture was stirred for another 3 h at −40 °C. The reaction was quenched with aqueous ammonium chloride (10 mL) at −78 °C and then warmed slowly to room temperature. The mixture was extracted with EtOAc (2 × 20 mL), and the combined organic layers were washed with 5% sodium bicarbonate, brine, dried, and concentrated. The crude product was purified via chromatography eluting with petroleum ether:ether (7:1) to give 1.4 g (88%) of the olefin.

Reference: Wee, A. G. H.; Yu, Q. *J. Org. Chem.* **2001**, *66*, 8935–8943.

#### 5.1.9.1.2 Valinol-Derived Oxazolidinone

A solution of diisopropylamine (3.42 mL, 24.4 mmol) in THF (30 mL) was cooled to −30 °C and *n*-BuLi in hexanes (2.5 M, 8.45 mL, 21.12 mmol) was added. After stirring the mixture for 45 minutes, a solution of the oxazolidinone (3.69 g, 16.25 mmol) in THF (20 mL) was added at −78 °C. The mixture was stirred at −78 °C for 1 h and iodomethane (5.06 mL, 81.3 mmol) was then added. After stirring for 30 min, the solution was allowed to reach room temperature. The mixture was extracted with ether, and the organic layer extract was washed with saturated ammonium chloride solution, saturated sodium bicarbonate, and brine. The dried solution (sodium sulfate) was concentrated in vacuo, and the residue was purified using flash chromatography eluting with 5% EtOAc and then 10% EtOAc in hexanes. This material was then crystallized from hexanes at −15 °C to deliver 2.47 g (63%) of the oxazolidinone as colorless crystals.

Reference: Guerlavais, V.; Carroll, P. J.; Joullié, M. M. *Tetrahedron: Asymmetry* **2002**, *13*, 675–680.

### 5.1.9.2 Myers' Asymmetric Alkylation

The Myers' asymmetric alkylation is a reaction between the enolate of a pseudoephedrine amide and an alkyl iodide in the presence of lithium chloride to give α-substituted amides with high diastereoselectivity. It has been noted that the lithium chloride must be anhydrous and flame-dried immediately before use. The α-substituted amides can then be transformed to the corresponding carboxylic acids (acid or base hydrolysis), alcohols (reduction with $LiH_2NBH_3$), aldehydes [reduction with $LiAlH(OEt_3)_3$], and ketones (nucleophilic attack by alkyllithium reagents). Review: Myers, A. G.; Yang, B. H.; Chen, H.; McKinstry, L.; Kopecky, D. J.; Gleason, J. L. *J. Am. Chem. Soc.* **1997**, *119*, 6496–6511.

A solution of *n*-butyllithium (2.50 M in hexanes, 27.96 mL, 69.9 mmol) was added via a cannula to a suspension of lithium chloride (9.39 g, 222 mmol) and diisopropylamine (10.6 mL, 75.3 mmol) in THF (50 mL) at −78 °C. The resulting suspension was warmed to 0 °C briefly and then cooled to −78 °C. An ice-cooled solution of the amide (8.12 g, 36.7 mmol) in THF (100 mL, followed by a 4 mL rinse) was added via a cannula. The mixture was stirred at −78 °C for 1 h, at 0 °C for 15 min at 23 °C for 5 min. The mixture was cooled to 0 °C and the iodide (5.08 g, 17.5 mmol) was added neat to the reaction via a cannula. After being stirred for 18.5 h at 0 °C, the reaction mixture was treated with half-saturated aqueous ammonium chloride solution (180 mL), and the resulting mixture was extracted with EtOAc (4 × 100 mL). The combined organic extracts were dried over sodium sulfate, filtered, and concentrated under reduced pressure. The residue was purified by flash chromatography eluting with ether:hexanes (65:35) to afford 6.52 g (97%) of the amide as a white solid.

Reference: Vong, B. G.; Abraham, S.; Xiang, A. X.; Theodorakis, E. A. *Org. Lett.* **2003**, *5*, 1617–1620.

### 5.1.10 Friedel–Crafts Reaction

The Friedel–Crafts reaction includes both alkylation and acylation although here only acylation is exemplified. For the Friedel–Crafts acylation, the electrophile could be either an acid chloride or an anhydride. The catalyst could be either a Lewis or protic acid.

Review: Bandini, M.; Melloni, A.; Ronchi-Umami, A. *Angew. Chem. Int. Ed.* **2004**, *43*, 550–556.

The anisole derivative (380 mg, 2 mmol) in dry $CH_2Cl_2$ (5 mL) under a nitrogen atmosphere was cooled to 0 °C. Anhydrous aluminum chloride (400 mg, 3 mmol) was added slowly and the mixture was stirred for 15 min. The acid chloride (456 mg, 2 mmol) in dry $CH_2Cl_2$ (5 mL) was added dropwise at 0 °C. The reaction mixture was stirred at 0 °C for 30 min and at room temperature overnight. After the reaction was deemed complete by TLC, the reaction mixture was poured on a mixture of crushed ice (5 g) and concentrated HCl (2 mL). The quenched reaction was stirred for 10 min and then extracted with $CH_2Cl_2$ (3 × 20 mL). The combined organic layers were washed with water (20 mL), brine (20 mL), and dried over sodium sulfate. The filtrate was then evaporated. The crude product was purified by column chromatography eluting with 20% EtOAc/petroleum ether to afford 535 mg (70%) of the ketone as a semi-solid.

Reference: Patil, M. L.; Borate, H. B.; Ponde, D. E.; Deshpande, V. H. *Tetrahedron* **2002**, *58*, 6615–6620.

### 5.1.11 Grignard Reaction (Organomagnesium Reagents)

The Grignard reaction involves an organomagnesium reagent reacting with an electrophile to form a new carbon–carbon bond. Organomagnesium reagents are readily prepared via several methods (see chapter 1). Magnesium reduction of an organohalide with magnesium turnings in the presence of an initiator such as iodine (*the initiator is necessary to remove the layer of magnesium oxide*) is the classical method (see chapter 1 for this method). However, in the past few decades halide–magnesium exchange has gained in popularity by allowing an organohalide to react with *i*-PrMgCl or *i*-PrMgBr and form the new organomagnesium reagent at low temperature (< 0 °C). Also, it is sometimes convenient to form these reagents via a metal–metal exchange by treating an organolithium reagent with $MgBr_2·Et_2O$ or $MgCl_2·Et_2O$ at low temperature (−70 °C). When forming the Grignard reagent, it is paramount that an ethereal solvent such as diethyl ether or THF is used because these solvents help stabilize these species via chelation. A variety of simple and functionalized aryl, heteroaryl, alkynyl, alkenyl, and alkylmagnesium reagents have been prepared. These reagents react readily with a host of electrophiles such as aldehydes, ketones, epoxides, allyl halides, etc. Reviews: (a) Knochel, P.; Dohle, W.; Gommerman, N.; Kneisel, F. F.; Kopp, F.; Korn, T.; Sapountzis, I.; Vu, V. A. *Angew. Chem. Int. Ed.* **2003**, *42*, 4302–4320. (b) Franzén, R. G. *Tetrahedron* **2000**, *56*, 685–691. (c) Wakefield, B. J. *Organomagnesium Methods in Organic Synthesis*; Academic Press: San Diego, 1995. (d) Ashby, E. C.; Laemmle, J.; Neumann, H. M. *Acc. Chem. Res.* **1974**, *7*, 272–280.

3-Triethylsilyloxy-1-iodo propene (5.56 g, 18.6 mmol) was dissolved in diethyl ether (50 mL) and cooled to −78 °C, and *tert*-BuLi (1.7 M in pentane, 19.5 mL, 33.1 mmol) was added dropwise. After 15 min at −78 °C, freshly prepared $MgBr_2 \cdot OEt_2$ (20.0 mL of a 0.83 M solution in 5:1 ether/benzene) was added via a cannula and stirring was continued at this temperature for 15 min. The amide (1.05 g, 4.14 mmol) was added as a solution in ether (30 mL) via a cannula followed by an ether rinse (3 mL). After 30 min at −78 °C, the cold bath was replaced with an ice bath, and the mixture was stirred for 2 h at 0 °C. The reaction mixture was cannulated into a vigorously stirred biphasic mixture of 0.5 N HCl (100 mL) and $CH_2Cl_2$ (100 mL) that was cooled to 0 °C. (*Note: HCl is required for this quench, as a milder acid such as ammonium chloride resulted in significant 1,4-addition of liberated N,O-dimethylhydroxylamine to the enone.*) The resulting mixture was extracted with $CH_2Cl_2$ (3 × 100 mL), and the combined extracts were washed with dilute aqueous sodium bicarbonate (100 mL), dried over magnesium sulfate, and concentrated. The crude product was purified by chromatography eluting with hexanes/EtOAc (9:1) to afford 1.35 g (90%) of the enone as a clear colorless oil.

Reference: Vanderwal, C. D.; Vosburg, D. A. Weiler, S.; Sorensen, E. J. *J. Am. Chem. Soc.* **2003**, *125*, 5393–5407.

### 5.1.12 Heck Coupling

The palladium catalyzed reaction between an aryl-X (X = I, Br, Cl, OTf, OTs, $N_2^+$) and an olefin is referred as the Heck reaction. In the traditional Heck reaction, the most common palladium catalysts include $Pd(OAc)_2$, $PdCl_2$, $Pd(PPh_3)_4$, and $Pd_2(dba)_3$; however, in the past 20 years, the literature has been enriched with a number of ligands that have improved the yields of the Heck reaction, and in numerous cases made the reaction possible. In situations where less reactive organohalides undergo the Heck sluggishly or not at all (i.e., arylchloride, electron-rich arylhalide), these electron-rich ligands accelerate the oxidative addition step with the organohalide. The types of ligands include monodenate phosphines, chelating diphosphines, phosphites, and palladacycles, to name a few. The range of olefinic coupling partners is also broad which includes α,β-unsaturated esters, α,β-unsaturated nitriles, styrene derivatives, isolated olefins, etc. The Heck reaction has been successfully used inter- and intramolecularly. Reviews: (a) Gurry, P. J.; Kiely, D. *Curr. Org. Chem.* **2004**, *8*, 781–794. (b) Dounay, A. B.; Overman, L. E. *Chem. Rev.* **2003**, *103*, 2945–2964. (c) Link, J. T. *Org. React.* **2002**, *60*, 157–534. (d) Whitcombe, N. J.; Hii, K. K. Gibson, S. E. *Tetrahedron* **2001**, *57*, 7431–7574. (e) Beletskaya, I. P.; Cheprakov, A. V. *Chem. Rev.* **2000**, *100*, 3009–3066. (f) Shibasaki, M.; Boden, C. D. J.; Kojma, A. *Tetrahedron* **1997**, *53*, 7371–7395. (g) Cabri, W.; Candiani, I. *Acc. Chem. Res.* **1995**, *28*, 2–7. (h) de Meijere, A; Meyer, F. E. *Angew. Chem. Int. Ed.* **1994**, *33*, 2379–2411. (i) Heck, R. F. *Org. React.* **1982**, *27*, 345–390. (j) Heck, R. F. *Acc. Chem. Res.* **1979**, *12*, 146–151.

*5.1.12.1 Standard Heck*

2-Chloro-5-bromonitro-benzene (30.0 g, 127 mmol), Pd(OAc)$_2$ (285 mg, 1.27 mmol), and triphenylphosphine (666 mg, 2.54 mmol) in dimethylformamide (360 mL) were added to triethylamine (24.7 mL, 178 mmol) and ethyl acrylate (138 mL, 1.27 mol). The reaction was stirred at 87 °C for 10 h, cooled to room temperature, and poured into a separatory funnel containing toluene (300 mL). The mixture was washed with 1 N HCl (300 mL) and water (2 × 200 mL). The organic extracts were concentrated to an oil that was crystallized in hexanes (60 mL). The solid was filtered to afford 30.8 g (95%) of the ester.

Reference: Caron, S.; Vazquez, E; Stevens, R. W.; Nakao, K.; Koike, H.; Murata, Y. *J. Org. Chem.* **2003**, *68*, 4104–4107.

### 5.1.12.2 Jeffrey's Ligandless Conditions

To a mixture of the vinyl iodide (580 mg, 0.858 mmol), potassium formate (217 mg, 2.57 mmol) and tetra-*n*-butylammonium bromide (277 mg, 0.858 mmol) in dimethylformamide (DMF, 34 mL) was added Pd(OAc)$_2$ (9.6 mg, 0.043 mmol) at room temperature. The reaction was stirred for 24 h at room temperature in the dark. After TLC indicated the disappearance of starting material, the reaction mixture was poured into water and extracted with hexanes (3 × 60 mL). The combined organic extracts were dried (MgSO$_4$), filtered, concentrated, and purified by flash chromatography to give 373 mg (79%) of the diene as a colorless oil.

Reference: Lee, K.; Cha, J. K. *J. Am. Chem. Soc.* **2001**, *123*, 5590–5591.

### 5.1.12.3 Heck Reactions of Aryl Chlorides (Fu Modification)

The Heck reaction of aryl chlorides tends to require reaction temperatures that exceed 120 °C. The aryl chloride lacks reactivity relative to an arylbromide or aryliodide because of the bond dissociation energy. The bond strength order is sp$^2$C–Cl > sp$^2$C–Br > sp$^2$C–I. The Fu modification decreases the reaction temperature significantly, and with activated chlorides the reaction can be run at room temperature in the presence of Pd(P(*t*-Bu)$_3$)$_2$ or Pd$_2$(dba)$_3$ and *N*-methyldicyclohexylamine (Cy$_2$NMe).

An oven-dried, 250-mL, three-necked, round-bottomed flask equipped with a reflux condenser (fitted with an argon inlet adapter), rubber septum, glass stopper, and a Teflon-coated magnetic stir bar was cooled to room temperature under a flow of argon. The flask was charged with (Pd(P(*t*-Bu)$_3$)$_2$) (0.482 g, 0.943 mmol, 3.0 mol% Pd) and again purged with argon. Toluene (32 mL) was added, and the mixture was stirred at room temperature, resulting in a homogeneous brown–orange solution. Chlorobenzene (3.20 mL, 31.5 mmol), Cy$_2$NMe (7.50 mL, 35.0 mmol), and butyl methacrylate (5.50 mL, 34.6 mmol) were then added successively via a syringe. The resulting mixture was allowed to stir at room temperature for 5 min, resulting in a homogeneous light-orange solution. The rubber septum was then replaced with a glass stopper, and the flask was heated in a 100°C oil bath under a positive pressure of argon for 22 h. On heating, the solution became a bright canary-yellow in color, and within 10–15 min the formation of a white precipitate (the amine hydrochloride salt) was observed. On completion of the reaction, shiny deposits of palladium metal formed on the sides of the flask, and a large quantity of white precipitate was present. The reaction mixture was allowed to cool to room temperature and then diluted with 100 mL of diethyl ether. The resulting solution was washed with 100 mL of H$_2$O, and the aqueous layer was extracted with three 50-mL portions of diethyl ether. The combined organic phases were washed with 100 mL of brine and then concentrated by rotary evaporation. Any residual solvent was removed at 0.5 mm. The crude product, a dark-brown oil, was then purified by flash column chromatography to afford 6.67–6.72 g (95%) of the α,β-unsaturated ester as a pale red–orange liquid. This liquid appears to be pure by $^1$H and $^{13}$C NMR spectroscopy; however, if desired, the discoloration can be removed by filtering the product through a small column of silica gel (3 cm diameter × 10 cm height) which furnishes 6.49–6.62 g (95–96%) of α,β-unsaturated ester as a clear colorless liquid.

Reference: Littke, A. F.; Fu, G. C. *Org. Syn.* **2004**, *81*, 63–70.

### 5.1.13 Henry Reaction (Nitro Aldol)

The Henry reaction or the nitroaldol is a classical reaction where the α-anion of an alkylnitro compound reacts with an aldehyde or ketone to form a β-nitroalcohol adduct. Over the decades, the Henry reaction has been used to synthesize natural products and pharmaceutical intermediates. In addition, asymmetric catalysis has allowed this venerable reaction to contribute to a plethora of stereoselective aldol condensations. Reviews: (a) Ballini, R.; Bosica, G.; Fiorini, D.; Palmieri, A. *Front. Nat.Prod. Chem.* **2005**, *1*, 37-41. (b) Ono, N. In *The Nitro Group in Organic Synthesis*; Wiley-VCH: Weinheim, 2001; Chapter 3: The Nitro-Aldol (Henry) Reaction, pp. 30-69. (c) Luzzio, F. A. *Tetrahedron* **2001**, *57*, 915–945.

To a solution of the aldehyde (480 mg, 2.44 mmol) and nitroethane (525 μL, 7.32 mmol) in *t*-BuOH/THF (1:1, 4.0 mL) was added *t*-BuOK (48 mg, 0.48 mmol) at room temperature. After 20 min, the mixture was diluted with ether (20 mL) and water (20 mL). The separated organic layer was washed with brine (20 mL), and the aqueous layers were back-extracted with ether (2 × 20 mL). The combined organic layers were dried (MgSO$_4$), filtered, and concentrated. The residue was purified by chromatography (silica gel) eluting with hexane:EtOAc (4:1) to afford 500 mg (76%) of the aldol adduct as a pale yellow oil which constituted an inseparable mixture of diastereomers.

Reference: Denmark, S. E.; Gomez, L. *J. Org. Chem.* **2003**, *68*, 8015–8024.

### 5.1.14 Hiyama Cross-Coupling Reaction

The palladium cross-coupling reaction between an aryl or vinylsilane with an organohalide in the presence of a fluoride source is referred to as the Hiyama reaction. Fluoride sources such as TASF [(Et$_2$N)$_3$S$^+$(Me$_3$SiF$_2$)$^-$], TBAF (tetrabutylammonium fluoride), KF, and CsF activate the silane towards metallation (without these activators, reactions usually do not occur). This reaction is tolerant of a wide array of functionality, including aldehydes, ketones, and esters. Coupling partners include arylhalides and vinylhalides. The silane nucleophile has been extended to include aryl- and vinylsilanols as well as aryltrimethoxysilanes. Reviews: (a) Denmark, S. E.; Ober, M. H. *Aldrichimica Acta* **2003**, *36*, 76–85. (b) Hiyama, T. In *Handbook of Organopalladium Chemistry for Organic Synthesis*; Negishi, E.; deMeijere, A., Eds.; Wiley-Interscience: New York, 2002; Chapter III.2.4: Overview of Other Palladium-Catalyzed Cross-Coupling Protocols, pp. 285–301. (c) Hiyama, T. *J. Organomet. Chem.* **2002**, *653*, 58–61. (d) Hiyama, T.; Shirakawa, E.. In *Topics in Current Chemistry: Cross-coupling Reactions A Practical Guide*. Miyaura, N., Ed; Springer-Verlag: Berlin, 2002; Vol. 219, Organosilicon Compounds, pp. 61–85. (e) Hiyama, T. In *Metal-Catalyzed Cross-coupling Reactions*; Diedrich, F.; Stang, P. J., Eds.; Wiley-VCH: New York, 1998; Chapter 10: Organosilicon Compounds in Cross-Coupling Reactions, pp. 421–453.

#### 5.1.14.1 Standard Hiyama

To a suspension of potassium fluoride (350 mg, 6.0 mmol, spray dried) in DMF (50 mL) were added ethyldifluoro-(4-methyl)silane (560 mg, 3.0 mmol), 3-iodobenzyl alcohol (250 mg, 2.0 mmol), and (η$^3$-C$_3$H$_5$PdCl)$_2$ (37 mg, 0.1 mmol). The resulting mixture was stirred at 100 °C for 15 h, cooled to room temperature, poured into saturated aqueous sodium bicarbonate solution, and extracted with diethyl ether (3 × 20 mL). The combined ethereal layer was then dried over magnesium sulfate. The solvent was removed under reduced pressure. The crude material was purified by silica gel chromatography eluting a mixture of hexane:EtOAc (5:1) to give 340 mg (86%) of 3-hydroxymethyl-4´-methylbiphenyl as a colorless solid.

Reference: Hatanaka, Y; Goda, K.; Okahara, Y.; Hiyama, T. *Tetrahedron* **1994**, *50*, 8301–8316.

### 5.1.14.2 Denmark Modification

Tetrabutylammonium fluoride (631 mg, 2.34 mmol) was dissolved in dry THF (2 mL) at room temperature under an atmosphere of dry nitrogen. The silanol (201 mg, 1.17 mmol) was added neat and the mixture was stirred for 10 min at room temperature. 4´-iodoacetophenone (246 mg, 1.0 mmol) was added to the mixture followed by Pd(dba)$_2$ (29 mg). The reaction mixture was stirred at room temperature for 10 min. The mixture was then filtered through a plug silica gel eluting with ether (100 mL).The solvent was removed in vacuo. The resulting residue was then purified by reverse phase chromatography (RPC18) eluting with methanol/H$_2$O (9/1) followed by Kugelrohr distillation to afford 201 mg (93%) of (*E*)-alkene as a colorless oil.

Reference: Denmark, S. E.; Wehrli, D. *Org. Lett.* **2000**, *2*, 565–568.

### 5.1.14.3 Fu Modification (Room Temperature Coupling)

In the air, PdBr$_2$ (10.6 mg, 0.040 mmol) and [HP(*t*-Bu)$_2$Me]BF$_4$ (24.8 mg, 0.10 mmol) were placed in a vial equipped with a stir bar. The vial was closed with a septum screwcap, and evacuated and refilled with argon three times. Tetrahydrofuran (2.4 mL) and tetrabutylammnoium fluoride [0.10 mL of 1.0 M (typically in THF), 0.10 mmol] were added, and the resulting mixture was stirred vigorously for 30 min at room temperature. To the resulting homogeneous orange–yellow solution was added (MeO)$_3$SiPh (0.225 mL, 1.20 mmol), tetrabutylammonium fluoride [2.3 mL of 1.0 M (typically in THF), 2.3 mmol), and the alkyl bromide (264 mg, 1.00 mmol). The reaction mixture was stirred vigorously for 14 h at room temperature. The reaction mixture was then filtered through a short pad of silica gel (washed with ~ 100 mL of ether or EtOAc) and concentrated. The crude product was purified by chromatography eluting with EtOAc:hexane (25:75) to deliver 172 mg (66%) of the morpholine amide as a colorless oil.

Reference: Lee, J.-Y.; Fu, G. C. *J. Am. Chem. Soc.* **2003**, *125*, 5616–5617.

## 5.1.15 Keck Stereoselective Allylation

The stannane reagents involved in the Keck stereoselective allylation are toxic and caution must be exercised in dealing with them. The catalyst is comprised of 1 equivalent of Ti(O$i$-Pr)$_4$ and 2 equivalents of ($R$)-(+)-BINOL. There are a variety of ways to generate the Ti-BINOL catalyst, and the enantioselectivity varies depending on the method of generation (see Keck, G. E.; Tarbet, K. H.; Geraci, L. S. *J. Am. Chem. Soc.* **1993**, *115*, 8467–8468 and Keck, G. E.; Krishnamurthy, D.; Grier M. C. *J. Org. Chem.* **1993**, *58*, 6543-6544). Reviews: (a) Gung, B. W. *Org. React.* **2004**, *64*, 1–113. (b) Denmark, S. E.; Almstead, N. G. In Modern Carbonyl Chemistry; Otera, J, Ed.; Wiley-VCH: Weinheim, 2000; Chapter 10: Allylation of Carbonyls: Methodology and Stereochemistry, pp. 299–402. (c) Chemler, S. R.; Roush, W. R. In Modern Carbonyl Chemistry. Otera, J.; Wiley-VCH: Weinheim, 2000; Chapter 11: Recent Applications of the Allylation Reaction to the Synthesis of Natural Products, pp. 403–490. (d) Denmark, S. E.; Fu, J. *Chem. Rev.* **2003**, *103*, 2763–2793.

Under argon, to a solution of ($R$)-(+)-BINOL (0.132 g, 0.46 mmol) over 4 Å molecular sieves (0.732 g) in methylene chloride (2.8 mL) was added Ti(O$i$-Pr)$_4$ (0.069 mL, 0.23 mmol). The resulting deep solution was stirred at ambient temperature for 1.5 h then treated with a solution of the aldehyde (0.5 g, 2.9 mmol) in methylene chloride (1.3 mL). The solution was stirred for 2 min, cooled to 0 °C, treated with the allylstannane (1.15 g, 3.2 mmol) over 2 min, and maintained at this temperature for 4.5 h. The reaction was then quenched with saturated sodium bicarbonate, extracted with EtOAc (3×), dried over magnesium sulfate, and concentrated in vacuo. The crude product was purified via flash chromatography using EtOAc:hexanes (1:9) to afford 0.495 g (70%) of the *anti*-alcohol and 28 mg (4%) of the *syn*-alcohol as colorless oils.

Reference: Smith, III, A. B.; Doughty, V. A.; Sfouggatakis, C.; Bennett, C. S.; Koyanagi, J.; Takeuchi, M. *Org. Lett.* **2002**, *4*, 783–786.

## 5.1.16 Kumada Coupling

The palladium catalyzed reaction between an organomagnesium reagent (Grignard reagent) and organohalide or organotriflate is the Kumada coupling. Although the Kumada coupling has been applied to a variety of coupling partners, the nucleophilic nature of the organomagnesium reagent deems it incompatible with susceptible functionality such as aldehydes, ketones, esters, epoxides, etc. Reviews: (a) Knochel, P.; Krosovskiy; Sapountzis, I. In *Handbook of Functionalized Organometallics*. Knochel, P. Ed.; Wiley-VCH: Weinheim, 2005; Vol. 1, pp. 155–158. (b) Hassan, J.; Sevignon, M; Gozzi, C.; Schulz, E.; Lemaire, M. *Chem. Rev.* **2002**, *102*, 1359.

## 5.1.16.1 With a Palladium Catalyst

[Scheme: vinyl iodide dihydropyran + CH2=CHMgBr, Pd(PPh3)4, PhH, THF, 70 °C → diene dihydropyran, 88%]

To the vinyl iodide (1.442 g, 4.938 mmol) in benzene (100 mL) was added vinyl magnesium bromide (1.0 M in THF, 19.75 mL, 19.75 mmol) and tetrakis(triphenylphosphine)palladium (286 mg, 0.247 mmol). *Note: degassing of the solvent is usually recommended for palladium cross-coupling reactions.* This reaction mixture was heated to 60–70 °C for 30 min, diluted with hexanes, and filtered through a pad of silica. After evaporation of the solvents, bulb to bulb Kugelrohr distillation provided 834 mg (88%) of the triene as a colorless oil.

Reference: Liu, P.; Jacobsen, E. N. *J. Am. Chem. Soc.* **2001**, *123*, 10772–10773.

## 5.1.16.2 With a Nickel Catalyst

[Scheme: naphthyl triflate with pyrrolopyridine substituent + MeMgBr, NiBr2(PPh3)2, Et2O, reflux → methyl-substituted naphthalene, 85%]

To a solution of the triflate (1.08 g, 2.65 mmol) in ether (17 mL) was added NiBr$_2$(PPh$_3$)$_2$ (59 mg, 0.080 mmol) followed by methylmagnesium bromide (3.0 M in ether, 2.2 mL, 6.6 mmol). The mixture was then refluxed for 24 h. The reaction mixture was quenched with water (30 mL) and diluted with CH$_2$Cl$_2$ (30 mL). *Note: it is advisable to cool the reaction mixture to 0 °C as an excess of the Grignard reagent is used.* The phases were separated, and the aqueous layer was extracted with CH$_2$Cl$_2$. The combined organic extracts were dried (MgSO$_4$), filtered, and evaporated in vacuo. The residue was purified by flash chromatography eluting with EtOAc to EtOAc:triethylamine (98:2) to give 618 mg (85%) of the pyridine as a white solid.

Reference: Spivey, A. C.; Fekner, T.; Spey, S. E.; Adams, H. *J. Org. Chem.* **1999**, *64*, 9430–9443.

## 5.1.17 Negishi Coupling

The palladium catalyzed reaction between an organozinc reagent and an organohalide, or organotriflate is referred as the Negishi coupling. Reviews: (a) Negishi, E,.; Hu, Q.; Huang, Z.; Qian, M.; Wang, G. *Aldrichimica Acta* **2005**, *38*, 71–88. (b) Knochel, P.; Leuser, H.; Gong, L.-Z.; Perrone, S.; Kneisel, F. F. In *Handbook of Functionalized Organometallics*. Knochel, P. Ed. Wiley-VCH: Weinheim, 2005; Vol. 1, Chapter 7: Polyfunctional Zinc Organometallics for Organic Synthesis, pp. 316–325. (c) Lessene, G. *Aust. J. Chem.* **2004**, *57*, 107–117. (d) Negishi, E. In *Handbook of Organopalladium Chemistry for Organic Synthesis*. Negishi, E.; deMeijere, A.,

Eds.; Wiley-Interscience: New York, 2002; Chapter III.2: Palladium-Catalyzed Carbon–Carbon Cross-Coupling, pp. 229–247.

### 5.1.17.1 Standard

4-Iodo-2-fluoropyridine (15.52 g, 69.60 mmol) was dissolved in dry THF (150 mL) and cooled to −70 °C. Then n-BuLi (73.08 mmol, typically a solution in hexanes) was added, and the reaction was stirred at −70 °C for 20 min. Subsequently, dry zinc chloride (10.44 g, 76.61 mmol) was added as a solution in dry THF (60 mL) while keeping the temperature below −60 °C. (*Use an internal thermometer.*) The reaction mixture was then warmed to room temperature whereupon tetrakis (triphenylphosphine)palladium (0.40 g, 0.35 mmol) and 2,4-dichloropyrimidine (7.26 g, 48.73 mmol) in THF (100 mL) were added, and the reaction mixture was refluxed until complete conversion. The reaction mixture was poured into a 10% aqueous EDTA solution and extracted with $CH_2Cl_2$. The crude material was purified by flash chromatography eluting with petroleum ether:EtOAc (10:1) to deliver 9.17 g (90%) of the pyrimidine as beige crystals.

Reference: Stanetty, P.; Hattinger, G.; Schnürch, M.; Mihovilovic, M. D. *J. Org. Chem.* **2005**, *70*, 5215–5220.

### 5.1.17.2 Fu Modification

Under argon, $ZnCl_2$ (0.5 M in THF, 3.15 mL, 1.6 mmol) was added by syringe to a Schlenk tube. *o*-Tolylmagnesium chloride (1.0 M in THF, 1.5 mL, 1.5 mmol) was then added dropwise, and the resulting mixture was stirred at room temperature for 20 min. NMP (2.2 mL) was added by syringe. After 5 min, $Pd(P(t-Bu)_3)_2$ (10.2 mg, 0.020 mmol) and 1-chloro-4-nitrobenzene (158 mg, 1.0 mmol) were added. The Schlenk tube was closed at the Teflon stopcock, and the reaction mixture was stirred in a 100 °C oil bath for 2 h. It was then allowed to cool to room temperature, and aqueous hydrochloric acid (1.0 M, 6 mL) was added. The resulting mixture was extracted with ether (4 × 8 mL), and the organic extracts were combined, washed with water (5 × 10 mL), dried ($MgSO_4$), and concentrated to afford a yellow solid. The crude product was purified via flash chromatography eluting with 3% ether in hexanes to furnish 200 mg (94%) of the biaryl as a pale-yellow solid.

Reference: Dai, C.; Fu, G. C. *J. Am. Chem. Soc.* **2001**, *123*, 2719–2724.

## 5.1.18 Nozaki–Hiyama–Kishi Reaction (Organochromium Reagents)

The formation of carbinols using organochromium reagents has come to be known as the Nozaki–Hiyama–Kishi reaction. A wide array of organochromium reagents have been generated, including alkenyl, alkynyl, allyl, aryl, and propargyl species. In general, they are formed from organohalides (usually bromides or iodides) and a mixture of $CrCl_2/NiCl_2$. The $NiCl_2$ is absolutely essential for reliable formation of these reagents. Organochromium reagents are chemoselective for aldehydes but do react with ketones to form carbinols. In addition to there chemoselectivity, these reagents have low basicity and are compatible with enolizable aldehydes and ketones, tolerate a multitude of functional groups, and add to aldehydes under mild conditions (*ambient temperature*).

Reviews: (a) Takai, K. *Org. React.* **2004,** *64*, 253–612. (b) Fürstner, A. *Chem. Rev.* **1999,** *99*, 991–1046. (c) Wessjohann, L. A.; Scheid, G. *Synthesis* **1999,** 1–36. (d) Cintas, P. *Synthesis* **1992,** 248–257.

A mixture of $NiCl_2$ (11.0 mg, 0.0849 mmol) and $CrCl_2$ (493 mg, 3.88 mmol) in degassed DMF (3 mL) was stirred at 0 °C for 10 min. A solution of the aldehyde (505 mg, 0.962 mmol) in DMF (3 mL) and a solution of the triflate (653 mg, 1.68 mmol) in DMF (3 mL) was added to the mixture at room temperature, and the resulting mixture was stirred at room temperature for 25 h. The reaction mixture was diluted with water and extracted with ether, washed with saturated sodium chloride, dried, and concentrated. The crude product was purified via chromatography eluting with *n*-hexane:EtOAc (5:1) to give 610.4 mg (83%) of a diastereomeric mixture of allylic alcohols.

Reference: Hirai, K.; Ooi, H.; Esumi, T.; Iwabuchi, Y.; Hatakeyama, S. *Org. Lett.* **2003,** *5*, 857–859.

## 5.1.19 Organocerium Reagent Addition to a Carbonyl

Organocerium reagents are generated in situ. They are typically synthesized by the reaction of an organolithium reagent or organomagnesium reagent and dry cerium chloride ($CeCl_3$). As $CeCl_3$ is commercially available as the heptahydrate, the water may be driven off by heating the finely ground solid at 140 °C at 0.5 torr. Cerium reagents offer several advantages over organolithium reagents. They are less basic and hence they work well when enolizable aldehydes or ketones are the coupling partners. Although they are less basic, the reagents are more nucleophilic. Also, organocerium reagents react preferentially in a 1,2-fashion with α,β-unsaturated carbonyl electrophiles. Review: Liu, H.-J.; Shia, K.-S.; Shang, X.; Zhu, B.-Y. *Tetrahedron* **1999,** *55*, 3803–3830.

**Scheme:**

Br–C₆H₄–OMOM (3-position) 

1. *n*-BuLi, THF, −78 °C
2. CeCl₃, −78 °C
3. TBSO–(CH₂)₄–CHO, −78 °C to rt
86%

→ TBSO–(CH₂)₄–CH(OH)–C₆H₄–OMOM (3-position)

A 250-mL, argon-flushed, three-necked flask was charged with anhydrous cerium chloride (9.90 g, 40.2 mmol) in dry THF (50 mL). The mixture was stirred for 4 h at room temperature and occasionally sonicated in an ultrasound cleaning bath (30 min) until the mixture became a fine white suspension. The flask was then cooled to −78 °C. In a separate 100-mL. argon-flushed, three-necked flask, 3-methoxymethylphenyl bromide (8.14 g, 37.5 mmol) was dissolved in dry THF (30 mL). After the solution was cooled to −78 °C, *n*-BuLi (1.6 M in hexanes, 23.5 mL, 37.6 mmol) was added via a syringe and the resulting brown solution was stirred for 15 min. The solution was then transferred to the flask containing the cerium chloride suspension by means of a transfer needle. The reaction mixture was stirred for 1 h at −78 °C upon which the solution turned an intense yellow. The aldehyde (5.75 g, 25 mmol) was added, and the mixture was stirred overnight while it was allowed to slowly come to room temperature. After the reaction was quenched with sodium bicarbonate (50 mL), the mixture was transferred to a separatory funnel and the layers were separated. The aqueous layer was extracted with $CH_2Cl_2$ (3 × 100 mL). The combined organic layers were washed with water and brine, dried over magnesium sulfate, filtered, and concentrated under reduced pressure. The crude product was purified by flash chromatography eluting with cyclohexane:EtOAc (6:1) to give 7.83 g (86%) of the alcohol as a colorless oil.

Reference: Kaiser, F.; Schwink, L.; Velder, J.; Schmalz, H.-G. *J. Org. Chem.* **2002**, *67*, 9248–9256.

### 5.1.20 Organolithium Reagents

Organolithium reagents have become one of the most important classes of organometallic compounds in modern organic synthesis for forming carbon–carbon bonds between aryllithiums, alkenyllithiums, alkynyllithium, alkyl lithiums, and an electrophile (i.e., aldehydes, ketones, esters, amides, epoxides, etc). Organolithium reagents are often prepared in situ from the appropriate arylhalide, alkenylhalide, etc., and a commercially available organolithium (i.e., *n*-BuLi, *sec*-BuLi, *tert*-BuLi) at low temperatures (i.e., −70 °C). A rapid halogen–metal exchange occurs to deliver the newly formed organolithium reagent and the butylhalide. The rate of the metal halogen exchange occurs as follows: R–I > R–Br >> R–Cl. The exchange is not trivial as the equilibrium process is sensitive to solvent (usually THF or diethyl ether), temperature, and rate of addition of the commercial lithium reagent. Lithium metal and dissolving metal reductions are used, but seen less frequently with the availability of the butyllithiums. Also, metallations of alkynes, aryl-, and heteroaryl ring systems are achieved with *n*-, *sec*-, and *tert*-butyllithium. Often a directing group is present with aryl compounds. Reviews: (a) Yus, M.; Foubelo, F. Polyfunctional Lithium

Organometallics for Organic Synthesis. In *Handbook of Functionalized Organometallics*; Knochel, P., Ed.; Wiley-VCH: Weinheim, 2005; Vol. 1, pp. 7–43. (b) Chincilla, R.; Nájera, C. *Tetrahedron* **2005**, *61*, 3139–3176. (c) Najera, C.; Yus, M. *Curr. Org. Chem.* **2003**, *7*, 867–926. (d) Sotomayor, N.; Lete, E. *Curr. Org. Chem.* **2003**, *7*, 275–300. (e) Nájera, C.; Sansano, J. M.; Yus, M. *Tetrahedron* **2003**, *59*, 9255–9303. (f) Wakefield, B. J. *Organolithium Methods*; Academic Press: San Diego (USA), 1988.

### 5.1.20.1 Halogen–Lithium exchange

#### 5.1.20.1.1 With n-BuLi

When using *n*-BuLi, Wurtz coupling is a possibility. This is a $S_N2$ reaction between the newly formed organolithium reagent and *n*-butylhalide (formed as a consequence of the metal–halogen exchange). See also organolithium generation in chapter 1.

To a solution of the arylbromide (2.58 g, 5.50 mmol) in dry THF (50 mL) was added *n*-BuLi (1.6 M in hexane, 3.44 mL, 5.50 mmol) at −78 °C under nitrogen. (*Note: the metal– halogen exchange process is exothermic so care is usually taken to add the organolithium reagent dropwise. It is also advisable to use an internal thermometer to closely monitor the reaction mixture temperature.*) The mixture was stirred for 10 min at −78 °C under nitrogen. Then, the aldehyde (1.19 g, 3.66 mmol) in dry THF (20 mL) was added, and the mixture was stirred at −78 °C for 1 h, warmed to 0 °C, and stirred for an additional 1 h. Saturated ammonium chloride was then added to quench the reaction. The solvent was removed under vacuum, and the resulting residue was dissolved in EtOAc, washed with 1 N HCl, saturated aqueous sodium bicarbonate, and brine (3×), and then dried over magnesium sulfate. After solvent removal, the crude product was purified by flash chromatography eluting with a gradient of hexane:EtOAc (100:1 to 20:1) to give 2.23 g (85%) of diastereomeric carbinols as a white semi-solid (the authors noted ~ 10% of impurities based on $^1$H NMR).

Reference: Deng, W.-P; Zhong, M.; Guo, X.-G. Kende, A. S. *J. Org. Chem.* **2003**, *68*, 7422–7427.

#### 5.1.20.1.2. With t-BuLi

Two equivalents of *t*-BuLi are used. One equivalent is required for the metal–halogen exchange, and the second equivalent reacts with the resulting *tert*-butyl halide to form isobutylene and a second equivalent of lithium halide.

A −78 °C solution of the vinyl iodide (1.05 g, 2.44 mmol) in dry THF (12 mL) was treated with *t*-BuLi (1.7 M in hexanes, 3.2 mL, 5.44 mmol). (*Note: the metal–halogen exchange process is exothermic so care is usually taken to add the organolithium reagent dropwise. It is also advisable to use an internal thermometer to closely monitor the reaction mixture temperature.*) After 2 min, the ketone (500 mg, 2.23 mmol) in THF (3 mL) was introduced via a syringe. The yellow mixture was stirred for 20 min, warmed to room temperature, quenched with saturated sodium bicarbonate solution, and extracted with ether. (*Note: it is prudent to cool the reaction mixture to 0 °C or lower prior to quenching even though these authors did not.*) The combined organic layers were washed with brine (2×), dried, and concentrated. The crude product was purified via chromatography eluting with 10% EtOAc in hexanes to afford 836 mg (70%) of the secondary alcohol as a colorless oil.

Reference: Paquette, L. A.; Montgomery, F. J.; Wang, T.-Z. *J. Org. Chem.* **1995**, *60*, 7857–7864.

## 5.1.20.2. Metallations

### 5.1.20.2.1 Alkynes

To a −78 °C solution of the alkyne (1.57 g, 7.92 mmol) in THF (25 mL) was added a solution of *n*-BuLi (2.5 M in hexanes, 3.33 mL, 8.32 mmol). The reaction mixture was stirred for 20 min and then a solution of the Weinreb amide (3.11 g, 8.71 mmol) in THF (5 mL) was added via a cannula. The reaction was allowed to warm to 0 °C and stirred for 2 h. The reaction mixture was then re-cooled to −78 °C and treated with 0.5 N HCl (5 mL). The mixture was allowed to warm to ambient temperature and stirred for 30 min. The mixture was diluted with ether and a saturated solution of sodium bicarbonate. The aqueous layer was separated and extracted with ether (3 × 30 mL). The combined ethereal extracts were washed with brine, dried (MgSO$_4$), filtered, and concentrated to dryness. The residue was purified by column chromatography eluting with hexanes:ether (20:1) to give 3.87 g (98%) of the ketone as a clear oil.

Reference: Roush, W. R.; Barda, D. A.; Limberakis, C.; Kunz, R. K. *Tetrahedron* **2002**, *58*, 6433–6454.

### 5.1.20.2.2. Ortho-metallations

Reviews: (a) Clark, R. D.; Jahangir, A. *Org. React.* **1995**, *47*, 1–314. (b) Snieckus, V. *Chem Rev.* **1990**, *90*, 879–933.

A solution of the naphthyloxazoline (200 mg, 0.79 mmol) in THF was cooled to −78°C and a solution of *n*-butyllithium (0.79 mL, 1.5 M in hexanes, 1.19 mmol) was added dropwise. The mixture was stirred at −78 °C for 2 h and then iodomethane (1.21 mL, 2.37 mmol) was added. The mixture was warmed to room temperature, stirred for 1 h, and then quenched with saturated aqueous ammonium chloride (30 mL). The mixture was extracted with $CH_2Cl_2$ (3 × 30 mL) and the combined organic extracts were dried over anhydrous sodium sulfate, filtered, and concentrated in vacuo. The crude material was purified by flash chromatography on silica gel to afford 259 mg (100%) of the dihydronaphthalene as a colorless oil.

*Aldehyde*

To a solution of the dihydronaphthalene (250 mg, 0.77 mmol) in $CH_2Cl_2$ (5 mL) was added methyl trifluoromethanesulfonate (227 mg, 1.38 mmol). The mixture was stirred at room temperature until the starting material had been completely consumed, as judged by TLC analysis (3 h). The mixture was cooled to 0 °C and a solution of $NaBH_4$ (111 mg, 2.92 mmol) in 4:1 MeOH:THF (3 mL) was slowly added. The mixture was warmed to room temperature and then quenched with saturated aqueous ammonium chloride (50 mL). The resulting mixture was extracted with $CH_2Cl_2$ (3 × 50 mL) and the combined organic extracts were dried over anhydrous sodium sulfate, filtered, and concentrated in vacuo. The resulting material was dissolved in 4:1 $THF/H_2O$ (5 mL), and oxalic acid (485 mg, 3.85 mmol) was added. The reaction mixture was stirred at room temperature for 12 h and then quenched with saturated aqueous sodium bicarbonate (50 mL). The resulting mixture was extracted with $CH_2Cl_2$ and the combined organic extracts were dried over anhydrous sodium sulfate, filtered, and concentrated in vacuo. The crude aldehyde was purified by flash chromatography on silica gel to afford 134 mg (76%) of the aldehyde as a colorless oil.

Reference: Rawson, D. J.; Meyers, A. I. *J. Org. Chem.* **1991**, *56*, 2292–2294.

### 5.1.21 Reformatsky Reaction (Organozinc)

The traditional Reformatsky reaction involves the conversion of a α-haloester to a α-organozinc ester in the presence of zinc metal and an initiator such as $I_2$ or 1,2-diiodoethane (*the initiator is necessary to remove the layer of zinc oxide*). The resulting organozinc reagent then reacts with an aldehyde or ketone to deliver a β-hydroxy ester. Other zinc sources such as Rieke zinc™, dissolving lithium reductions with zinc chloride, and many others have been reported. Over the past several decades this reaction has evolved to include other metals such as chromium, samarium iodide (Kagan reagent), and indium. In addition, the number of electrophiles has increased and some include nitriles, esters, lactones, epoxides, aziridines, and Michael acceptors. The Reformatsky reaction has been reported inter- and intramolecularly and used in asymmetric synthesis.

Reviews: (a) Ocampo, R.; Dolbier, Jr. W. R. *Tetrahedron* **2004,** *60,* 9325–9374. (b) Fürstner, A. *Synthesis* **1989,** 571–590.

Ethyl bromoacetate (1.14 mL, 10 mmol) was added slowly to a mixture of the aldehyde (2.29g, 6 mmol) and acid-washed zinc dust (0.78 g, 12 mmol) in dry THF (40 mL) at room temperature under a nitrogen atmosphere. The reaction mixture was stirred vigorously for 2 h before the addition of aqueous ammonium chloride (20 mL) and brine (20 mL). After filtration through Celite, the precipitate was washed well with EtOAc (4 × 10 mL) and the organic phase was separated. The aqueous phase was extracted with EtOAc (4 × 10 mL), the combined organic extracts were dried (magnesium sulfate), and the solvent was removed under reduced pressure to give an oily residue. The material was purified by flash chromatography, eluting with hexane:EtOAc (4:1), and delivered 1.92 g (72%) of the major diastereomer as a colorless oil followed by 512 mg (23%) of the minor diastereomer as a colorless oil for a total yield of 2.43 g (95%).

Reference: Ding, Y.; Wang, J.; Abboud, K. A.; Xu, Y.; Dolbier, Jr., W. R.; Richards, N. G. L. *J. Org. Chem.* **2001,** *66,* 6381–6388.

### 5.1.22 Roush Crotylation

While the Roush crotylboronate reagent takes a whole day to prepare, it works extremely well with regard to both yield and stereoselectivity, especially for the matched substrates.

Reviews: (a) Denmark, S. E.; Almstead, N. G. In *Modern Carbonyl Chemistry*; Otera, J., Ed.; Wiley-VCH: Weinheim: 2000; Chapter 10: Allylation of Carbonyls: Methodology and Stereochemistry, pp. 299–402. (b) Chemler, S. R.; Roush, W. R. In *Modern Carbonyl Chemistry*; Otera, J. Ed.; Wiley-VCH: Weinheim: 2000; Chapter 11: Recent Applications of the Allylation Reaction to the Synthesis of Natural Products, pp. 403–490. (c) Denmark, S. E.; Fu, J. *Chem. Rev.* **2003,** *103,* 2763–2793.

## CARBON–CARBON BOND FORMATION

A solution of (R,R)-diisopropyltartrate (E)-crotylboronate (46.0g, 0.152 mol) in dry toluene (1 L) was treated with powdered 4 Å molecular sieves (6 g) at −78 °C. A solution of the aldehyde (13.1 g, 0.076 mol) in dry toluene (50 mL) was cooled to −78 °C and added dropwise into the suspension via a cannula. The resultant mixture was stirred at −78 °C for 1 h before an aqueous solution of sodium hydroxide (2 N, 70 mL) was added. The mixture was warmed to 0 °C, stirred for 30 min, and filtered through a pad of Celite. The aqueous layer was extracted with ether (3 × 250 mL) and the combined organic phases were washed with brine (200 mL), dried over magnesium sulfate, filtered, and concentrated in vacuo. The crude product was purified via flash chromatography eluting with hexanes:EtOAc (7:1) to afford 15.2 g (88%, 90% d.e.) of the alcohol as a colorless oil.

Reference: Smith, III, A. B.; Zheng, J. *Tetrahedron* **2002**, *58*, 6455–6471.

### 5.1.23 Sakurai Reaction

The Sakurai allylation works with a variety of Lewis acid catalysts such as $TiCl_4$, $AlCl_3$, $SnCl_4$, $EtAlCl_2$, in addition to $BF_3 \cdot OEt$. Reviews: (a) Denmark, S. E.; Almstead, N. G. In *Modern Carbonyl Chemistry*; Otera, J., Ed.; Wiley-VCH: Weinheim, 2000; Chapter 10: Allylation of Carbonyls: Methodology and Stereochemistry, pp. 299–402. (b) Chemler, S. R.; Roush, W. R. In *Modern Carbonyl Chemistry*; Otera, J., Ed.; Wiley-VCH: Weinheim, 2000; Chapter 11: Recent Applications of the Allylation Reaction to the Synthesis of Natural Products, pp. 403–490. (c) Denmark, S. E.; Fu, J. *Chem. Rev.* **2003**, *103*, 2763–2793.

#### 5.1.23.1 Traditional

To the aldehyde (0.511 g, 1.5 mmol) in dry $CH_2Cl_2$ (15 mL) under argon was added allyltrimethylsilane (0.48 mL, 3 mmol) followed by boron trifluoride etherate (0.19 mL, 1.5 mmol) at −80 °C. The reaction was followed by TLC. After 5 h of stirring at −80 °C, the reaction was hydrolyzed with a 2:1 mixture of saturated aqueous ammonium chloride solution and ammonia (28% in water). The layers were separated, and the aqueous layer was extracted twice with $CH_2Cl_2$. The combined organic phases were dried over anhydrous magnesium sulfate and the solvents removed in vacuo. The solid obtained was purified by flash chromatography on silica gel eluting with cyclohexane:EtOAc: triethylamine (9:1:0.2) to afford 364 mg (63%) of the homoallylic alcohol as a white solid.

Reference: Bejjani, J.; Chemla, F.; Audouin, M. *J. Org. Chem.* **2003**, *68*, 9747–9752.

### 5.1.23.2 Denmark's Version Using a Chiral Bis-phosphoramide

To a solution of the bis-phosphoramide catalyst (59 mg, 0.1 mmol) in $CH_2Cl_2$ (1 mL) and diisopropylethylamine (1.0 mL) under $N_2$ at −78 °C was added allyltrichlorosilane (580 µL, 4.0 mmol). The solution was stirred at −78 °C for 10 min before benzaldehyde (200 mL, 2.0 mmol) was added. The resulting mixture was stirred at this temperature for 8 h whereupon the cold solution was poured in to a mixture of saturated aqueous $NaHCO_3$ (10 mL) and saturated aqueous KF (10 mL) at 0 °C with vigorous stirring. The mixture was stirred for 2 h at room temperature and then filtered through Celite. The layers were then separated and the aqueous layer was extracted with $CH_2Cl_2$ (3 × 30 mL). The combined organic extracts were dried ($MgSO_4$), filtered, and concentrated. The oily residue was purified by column chromatography (silica gel) eluting with $CH_2Cl_2$:pentane (3:1) followed by $CH_2Cl_2$ to give 254 mg (85%) of the benzylic alcohol in 87% ee.

Reference: Denmark, S. E.; Fu, J. *J. Am. Chem. Soc.* **2001**, *123*, 9488–9489.

### 5.1.24 Schwartz's Reagent

Schwartz hydrozirconation works on both alkenes and alkynes even at room temperature. It works significantly better in ether solvents than hydrocarbon solvents. In addition, the reaction intermediate can be transmetallated into other organometallics that can be used in palladium-catalyzed reactions or trapped with electrophiles. Review: Schwartz, J.; Labinger, J. A. *Angew. Chem. Int. Ed.* **1976**, *15*, 333–340.

To a stirred suspension of Schwartz's reagent, $Cp_2Zr(H)Cl$ (1.36 g, 5.26 mmol) in THF (15 mL) at room temperature was added a solution of the alkyne (0.585 g, 2.50 mmol) in THF (15 mL). The mixture was stirred for 24 h and then cooled to 0 °C whereupon a solution of iodine (1.27 g, 5.0 mmol) in THF (8 mL) was added dropwise. The resulting reaction mixture was stirred at room temperature for 30 min and then quenched with a dilute solution of sodium thiosulfate. The mixture was extracted with ether (3 ×), and the combined organic extracts were washed successively with a dilute solution of sodium thiosulfate and brine. After drying over anhydrous magnesium sulfate, the suspension was filtered, and the solvent was removed in vacuo to give a semi-solid residue that was extracted with pentane (3×). After removal of the pentane in vacuo, the crude product was chromatographed on silica gel eluting with pentane to give 0.732 g (81%) of the vinyl iodide as a yellow oil.

Reference: Organ, M. G.; Wang, J. *J. Org. Chem.* **2003**, *68*, 5568–5574.

### 5.1.25 Shapiro Reaction

The Shapiro reaction is the conversion of a ketone to an alkene via a tosylhydrazone. The tosylhydrazone is treated with two equivalents of strong base, typically *n*-BuLi or MeLi, to afford an alkenyllithium species that is quenched with an electrophile to form an alkene. See also section 4.6.6 Shapiro reaction. Reviews: (a) Chamberlin, A. R.; Bloom, S. H. *Org. React.* **1990**, *39*, 1–83. (b) Shapiro, R.H. *Org. React.,* **1976**, *23*, 405.

Tosylhydrazine (3.53g, 18.9 mmol) was added to 60% aqueous MeOH (46 mL) and then heated to 60 °C. The heptanone (2.33 g, 18.8 mmol) was then added dropwise to the clear solution. The reaction was then immediately stored at 5 °C for 15 h. The resultant white crystals were filtered, washed with 60% aqueous methanol, and air dried for 10 min. The hydrazone was obtained as white crystals (4.91 g, 89%).

A suspension of the tosylhydrazone (1.04 g, 3.59 mmol) and TMEDA (15 mL) was cooled to –78 °C under a nitrogen atmosphere. A solution of *n*-BuLi (2.3 M, 4.6 mL, 10.58 mmol) was added dropwise to the frozen suspension, and the resulting solution was kept at –78 °C for 15 min and then allowed to warm to ambient temperature whereupon it turned a dark red color. (*Caution: nitrogen is produced in this reaction and proper ventilation of the gas is required. Do not use a sealed system!*) This solution was stirred for a further 5 h. It was cooled to 0 °C, DMF (0.4 mL, 5.18 mmol) was added, and the solution stirred at ambient temperature overnight. It was then poured into a solution of hydrochloric acid (7.5%, 60 mL) and extracted with $CH_2Cl_2$ (4 × 40 mL). The combined organic phases were washed with saturated sodium chloride (40 mL) and dried. The solvent was then removed in vacuo to give ~2 g (~100%) of aldehyde.

Reference: Hamon, D. P. G.; Tuck, K. L. *J. Org. Chem.* **2000**, *65*, 7839–7846.

### 5.1.26 Sonogashira Coupling

The palladium catalyzed reaction between a terminal alkyne and an unsaturated organohalide or organotriflate in the presence of CuI and a tertiary organic amine is the

Sonogashira reaction. Unsaturated organic species include arylhalides (X = I, Br, Cl), aryltriflates, vinylhalides, and vinyltriflates. This is a reaction that also tolerates a large number of functional groups. Aryliodides typically react with terminal alkynes under Sonogashira conditions at room temperature; however, less reactive coupling partners such as arylbromides require elevated temperatures. Additives such as tetrabutylammonium bromide and potassium iodide accelerate the Songoashira reaction of less reactive coupling partners. Also, a significant amount of work has been conducted to identify ligands.

Review: Sonogashira, K. In *Handbook of Organopalladium Chemistry for Organic Synthesis*; Negishi, E.; deMeijere, A., Eds; Wiley-Interscience: New York, **2002**, Chapter III.2.8: Palladium-Catalyzed Alkynylation, pp. 493–535.

To a nitrogen degassed solution of triethylamine (3.1 mL, 22.2 mmol) in $CH_3CN$ (50 mL) were added the alkyne (2.6 g, 7.5 mmol), the aryl iodide (3.5 g, 7.4 mmol), $PdCl_2(PPh_3)_2$ (61.6 mg, 0.088 mmol), and CuI (16.8 mg, 0.088 mmol), and the mixture was stirred at 25 °C for 12 h. After removal of the solvent, the residue was filtered, concentrated, and purified by flash chromatography on silica gel eluting with petroleum ether:EtOAc (10:1) to give 4.48 g (90%) of the alkyne as a white solid.

Reference: Li, C. C.; Xie, Z. X.; Zhang, Y. D.; Chen, J. H.; Yang, Z. *J. Org. Chem.* **2003**, *68*, 8500–8504.

### 5.1.27 Stille Reaction

The palladium catalyzed coupling between an organostannane and organohalide or organotriflate is referred to as the Stille reaction. This reaction is tolerant of a variety of functional groups with a plethora of intra- and intermolecular reactions reported in the past 20 years. Consequently, an assortment of simple and multifunctionalized alkynyl, vinyl, aryl, allyl, benzyl, and alkylorganotin reagents have been synthesized. Also, a large number of diverse coupling partners, including acid chlorides, benzyl halides, allyl halides, allyl acetates, aryl halides, heteroaryl halides, aryltriflates, heteroaryltriflates, alkenyl halides, alkeynyl triflates, alkynyl halides, and alkyl halides have been used. The Sn reagents can be synthesized by a host of protocols; however, three of the most popular include palladium coupling between an organohalide (or triflate) and tetramethyltin, metal–halogen exchange of an organohalide followed by quenching of the resulting organolithium with trimethyltin chloride, and Pd-coupling of an organohalide (or triflate) with $Me_3SnSnMe_3$. Pd(0) and Pd (II) catalysts have been utilized for the Stille reaction, including $Pd(PPh_3)_4$, $Pd(PPh_3)_2Cl_2$, $BnPd(PPh_3)_2Cl$,

PdCl$_2$(MeCN)$_2$, Pd(OAc)$_2$, Pd(dba)$_2$, Pd$_2$(dba)$_3$/P(2-furyl)$_3$, and Pd(OAc)$_2$/P(o-tol)$_3$. Typically, the coupling is done in hot or refluxing solvent. Some solvents include THF, benzene, toluene, DMF, CH$_3$CN, dioxane, and NMP. The major disadvantage of the Stille reaction, compared with other organometallic palladium couplings, is the toxicity of the stannane reagents. Great care should be exercised in preventing exposure of the tin reagent via inhalation or skin absorption. In addition, proper disposal of the tin waste is pertinent. Removal of excess reagent during the workup and purification can also present difficulties. Reviews: (a) Fouquet, E.; Herve, A. In *Handbook of Functionalized Organometallics*. Knochel, P., Ed.; Wiley-VCH: Weinheim, 2005; Vol. 1, Chapter 6: Polyfunctional Tin Organometallics for Organic Synthesis, pp. 203–215. (b) Espinet, P. Echavarren, A. M. *Angew. Chem. Int. Ed.* **2004**, *43*, 4704–4734. (c) Fugami, K.; Kosugi, M. In *Topics in Current Chemitry: Cross-coupling Reactions A Practical Guide.* Miyaura, N., Ed.; Springer-Verlag: Berlin, 2002; Vol. 219, Organotin Compounds, pp. 87–130. (d) Kosug, M.; Fugami, K. In *Handbook of Organopalladium Chemistry for Organic Synthesis*; Negishi, E.; deMeijere, A., Eds.; Wiley-Interscience: New York, 2002; Chapter III. 2.3: Overview of the Stille Protocol with Sn, pp. 263–283. (e) Farina, V.; Krishnamurthy, V.; Scott, W. *J. Org. React.* **1997**, *50*, 1-652. (f) Stille, J. K. *Angew. Chem. Int. Ed.* **1986**, *25*, 508-524.

To a stirred suspension of the bromoindene (500 mg, 1.47 mmol) in toluene (20 mL) were added tetrakis(triphenylphosphine)palladium (150 mg, 0.09 equivalents) and then tributyl(vinyl)tin (0.5 mL, 1.15 equivalents), and the mixture was stirred and refluxed for 24 h. The reaction mixture was then cooled to room temperature, diluted with CH$_2$Cl$_2$ (20 mL), filtered through a pad of Celite, and evaporated to dryness. The resulting crude product was taken up in CH$_2$Cl$_2$ (100 mL), and aqueous saturated potassium fluoride solution (150 mL) was added. The mixture was then stirred vigorously overnight. The organic phase was then separated, and the aqueous layer was extracted with EtOAc (2 × 30 mL). The combined organic layers were dried over sodium sulfate, filtered, and evaporated to dryness. The crude product was dissolved in chloroform and purified by flash chromatography eluting with hexanes:EtOAc (7:3) to afford 314 mg (84%) of the 7-vinylindene as a white solid.

(*Note on workup: if the final product is soluble in acetonitrile but lacks solubility in hexanes, a solution of the crude product in acetonitrile can be washed several times with hexanes to extract the tin residues.*)

Reference: Hanessian, S.; Papeo, G.; Angiolini, M.; Fettis, K.; Beretta, M.; Munro, A. *J. Org. Chem.* **2003**, *68*, 7204–7218.

### 5.1.28 Stille–Kelly Reaction

The Stille–Kelly reaction is a variant of the Stille coupling where an organodihalide undergoes an intramolecular reaction in the presence of hexamethylditin and a palladium catalyst.

**Reaction scheme:** Diaryl isoxazole with two ortho-iodo substituents → phenanthro[1,2]oxazole, using Me₃SnSnMe₃, Pd(PPh₃)₂Cl₂, 1,4-dioxane, 115 °C, sealed tube, 89%.

A heavy wall-pressure tube was charged with the diarylisoxazole (418 mg, 0.89 mmol), Pd(PPh$_3$)$_2$Cl$_2$ (19.3 mg, 0.027 mmol), and degassed dioxane (19 mL) under nitrogen. A solution of Me$_6$Sn$_2$ (440 mg, 1.33 mmol) in degassed 1,4-dioxane (7.4 mL) was added dropwise to the resulting suspension, and after flushing with nitrogen at room temperature for 15 min, the mixture was heated at 115 °C in an autoclave for 50 min. After cooling, the resulting black suspension was centrifuged and the deposited black palladium was washed with CH$_2$Cl$_2$ (1 × 3 mL). The combined organic layers were washed with saturated potassium fluoride solution (1 × 7 mL), dried (Na$_2$SO$_4$), and the solvent was evaporated in vacuo. The resulting yellow residue was purified by flash chromatography eluting with hexane:EtOAc (30:70) providing a colorless oil which was crystallized from methanol to deliver 0.173 g (89%) of the phenanthro[1,2]oxazole as a colorless powder.

Reference: Olivera, R.; SanMartin, R.; Tellitu, I.; Domínguez, E. *Tetrahedron* **2002**, *58*, 3021–3037.

### 5.1.29 Suzuki Coupling (Suzuki–Miyaura)

The Pd catalyzed reaction between an organoboronic acid or organic boronic ester and an organohalide in the presence of a base is referred as the Suzuki or Suzuki–Miyaura reaction. The Suzuki reaction has become one of the most powerful carbon–carbon bond forming reactions in organic synthesis because of its tolerance of functionality, lack of toxicity associated with boronic acids, the relative ease of synthesizing boronic acids or boronic esters, and the large number of commercially available boronic acids. Despite these advantages, boronic acids are often contaminated with varying amounts of boronic anhydrides and base sensitive functionality might be compromised in the presence of the required Lewis base. In the traditional Suzuki reaction, a number of catalysts have been used, including Pd(PPh$_3$)$_4$, Pd(dppf)Cl$_2$, Pd(PPh$_3$)$_2$Cl$_2$, Pd(dba)$_2$/dppf, Pd(OAc)$_2$, Pd(OAc)$_2$/dppf, and Pd(OAc)$_2$/PPh$_3$. Over the years, much development has also been done on other ligands, including AsPh$_3$, P(*o*-tol)$_3$ and P(*t*-Bu)$_3$. However, the reader is encouraged to consult the reviews as the ligands are vast. In addition, nickel catalysts such as NiCl$_2$(PPh$_3$)$_2$ have been successful involving arylchlorides. Common bases include KF, CsF, NaHCO$_3$, Na$_2$CO$_3$, K$_2$CO$_3$, Cs$_2$CO$_3$, K$_3$PO$_4$, NaOH, Ba(OH)$_2$, NaOH, Tl$_2$CO$_3$ (*toxic*), TlOH (*toxic*), and triethylamine. Typically, the reactions are run above 70 °C in solvent systems that include THF/water, 1,4-dioxane/water, toluene/EtOH/H$_2$O, 1,2-dimethoxyethane (DME)/water, and DMF (anhydrous conditions).

Reviews: (a) Knochel, P.; Ila, H.; Korn, T. J.; Baron. O. Functionalized Organoborane Derivatives in Organic Synthesis. In *Handbook of Functionalized Organometallics*. Knochel, P. Ed.; Wiley-VCH: Weinheim, 2005; Vol. 1,, pp. 45–108. (b) Bai, L.; Wang, J.-X. *Curr. Org. Chem.* **2005**, *9*, 535-553. (c) Bellina, F.; Carpita, A.; Rossi, R. *Synthesis* **2004**, 2419-2440. (d) Kotha, S.; Lahiri, K.; Kashinath, D.

*Tetrahedron* **2002**, *58*, 9633-9695. (e) Knochel, P.; Ila, H.; Korn, T. J.; Baron, O. In *Handbook of Functionalized Organometallics*. Knochel, P. Ed.; Wiley-VCH: Weinheim, 2005; Chapter 3: Functionalized Organoborane Derivatives in Organic Synthesis, pp. 45–108. (f) Miyaura, N. In *Topics in Current Chemistry: Cross-coupling Reactions A Practical Guide*. Miyaura, N., ed.; Springer-Verlag: Berlin, 2002; Vol. 219, Organoboron Compounds, pp. 11–59. (g) Suzuki, A. *Handbook of Organopalladium Chemistry for Organic Synthesis*. Wiley-Interscience: New York, 2002; Vol 1, Chapter III.2.2: Overview of the Suzuki Protocal with B, pp. 249–262. (h) Chemler, S. R. Trauner, D; Danishefsky, S. J. *Angew. Chem. Int. Ed.* **2001**, *40*, 4544-4568. (i) Miyaura, N.; Suzuki, A. *Chem. Rev.* **1995**, *95*, 2457–2483. (j) Suzuki, A. *Acc. Chem. Res.* **1982**, *15*, 178-184.

### 5.1.29.1 Via Boronic Acids

#### 5.1.29.1.1 Formation of the Boronic Acid

Aryl bromide (1-Br-2-OBn-naphthalene) 
1. *n*-BuLi, Et$_2$O, –78 to 0 °C
2. B(OMe)$_3$, –78 °C to rt
→ Boronic acid (1-B(OH)$_2$-2-OBn-naphthalene)
83%

To a suspension of the aryl bromide (6.26 g, 20.0 mmol) in ether (75 mL) at –78 °C was added *n*-BuLi (2.5 M in hexanes, 8.0 mL, 20 mmol), and the mixture was stirred at 0 °C for 1 h. (*Note: typically the halogen metal exchange is done at –78 °C or lower to reduce the chance of side reactions while generating the organolithium. In this case, butyl bromide is produced which could react with the generated aryllithium. See chapter 1 for Wurtz coupling.*) After re-cooling to –78 °C, the mixture was treated with trimethyl borate (2.5 mL, 22 mmol) and allowed to warm to room temperature overnight. The resulting mixture was quenched with 1 M HCl (50 mL) and stirred at room temperature for 45 min. The phases were separated, and the extraction was completed with CH$_2$Cl$_2$. The combined organic extracts were dried (magnesium sulfate), filtered, and evaporated in vacuo to give 4.62 g (83%) of the boronic acid as a white powder.

Reference: Spivey, A. C.; Fekner, T.; Spey, S. E.; Adams, H. *J. Org. Chem.* **1999**, *64*, 9430–9443.

#### 5.1.29.1.2 Standard Suzuki

Pd(PPh$_3$)$_4$ can be an unreliable catalyst because it is air, light, and moisture sensitive; however, Pd(dppf)Cl$_2$·CH$_2$Cl$_2$ and Pd(PPh$_3$)$_2$Cl$_2$ enjoy greater stability and are excellent alternatives. All three catalysts are commercially available.

Boronic acid + Br-heteroaryl, Pd(PPh$_3$)$_4$, Na$_2$CO$_3$, PhMe, H$_2$O, EtOH, reflux → coupled product, 82%

To a solution of aryl bromide (1.49 g, 7.00 mmol) in toluene (28 mL) and ethanol (5 mL) was added an aqueous solution of sodium carbonate (2 M, 28 mL) followed by tetrakis-(triphenylphosphine) palladium (243 mg, 0.210 mmol) and aryl boronic acid (2.53 g, 9.1 mmol). The reaction mixture was then brought to reflux. *Note: degassing of the reaction mixture is usually recommended; however, it is not always necessary.* Additional portions of the boronic acid (680 mg, 2.44 mmol) and tetrakis(triphenylphosphine)palladium (81 mg, 0.070 mmol) were added at 90 min and 4.5 h. The reaction mixture was heated at reflux for a total of 24 h. The phases were separated, and the aqueous layer was extracted with $CH_2Cl_2$. The combined organic extracts were dried ($MgSO_4$) and evaporated in vacuo. The residue was purified by flash chromatography eluting with EtOAc to EtOAc:triethylamine (98:2) to give 2.09 g (82%) of the naphthalene as a white crystalline solid.

Reference: Spivey, A. C.; Fekner, T.; Spey, S. E.; Adams, H. *J. Org. Chem.* **1999**, *64*, 9430–9443.

*5.1.29.1.3 Fu Modification*

Open to the atmosphere, 4-bromoanisole (1.87 g, 10 mmol), *o*-tolylboronic acid (1.50 g, 11 mmol), KF (spray dried, dried in an oven overnight prior to use, 1.92 g, 33 mmol), and THF (10 mL) were added to a 100-ml round-bottomed Schlenk flask equipped with a stir bar. The reaction system was flushed with argon for about 5 min. P(*t*-Bu)$_3$ (1.9 × 10$^{-4}$ M stock solution in THF; 2.31 mL, 5.0 × 10$^{-5}$ mmol) and Pd$_2$(dba)$_3$ (2.16 × 10$^{-5}$ M stock solution in THF; 2.31 mL, 5.0 × 10$^{-5}$ mmol) in THF were added sequentially. After 48 h at room temperature, the reaction mixture was diluted with ether or EtOAc, filtered through a pad of silica gel with copious washings, and then concentrated. The crude product was then purified via column chromatography eluting with 5% ether in hexane to yield 1.94 g (98%) of 4-methoxy-2´-methyl-biphenyl as a colorless liquid.

Reference: Littke, A. F.; Dai, C.; Fu, G. C. *J. Am. Chem. Soc.* **2000**, *122*, 4020-4028.

*5.1.29.2 Via Pinacolatoboronic Ester*

*5.1.29.2.1a Formation of a Boronic Ester: Miyaura Protocol*

Useful solvents for the formation of the boronate ester include THF, 2-Methyltetrahydrofuran, 1,4-dioxane, dimethylformamide, and dimethylsulfoxide (DMSO). Pd(dppf)Cl$_2$ and potassium acetate are typically used for the formation of the pinacolatoboronic ester. Often after workup, the crude boronate is used for the subsequent Suzuki reaction. As can be seen with the example below, various functional groups tolerate this transformation.

Bispinacolatodiboron (0.855 g, 3.37 mmol), Pd(dppf)Cl$_2$·CH$_2$Cl$_2$ (0.500 g, 0.611 mmol), and potassium acetate (0.900 g, 9.18 mmol) were all added to a flask containing the bromide (2.31 g, 3.06 mmol). This vessel was then equipped with an air condenser, and degassed DMSO (20 mL, prepared by three iterations of the freeze–thaw method) was added at 25 °C. The reaction contents were then warmed to 85 °C and stirred for 6 h. On completion, the reaction mixture was diluted with EtOAc (50 mL), filtered through a short plug of silica, and extracted with EtOAc (3 × 50 mL). The combined organic layers were then washed with water (2 × 50 mL) and brine (50 mL), dried (MgSO$_4$), and concentrated. The resultant yellow–green residue was purified by flash chromatography (silica gel) eluting with EtOAc:hexanes (1:2 to 1:1) to give 1.99 g (81%) of boronic ester as a light yellow foam.

Reference: Nicolaou, K. C.; Snyder, S. A.; Giuseppone, N.; Huang, X.; Bella, M.; Reddy, M. V.; Rao, P. B.; Koumbis, A. E.; Giannakakou, P.; O'Brate, A. *J. Am. Chem. Soc.* **2004**, *126*, 10174–10182.

### 5.1.29.2.1 Formation of a Boronic Ester: Halogen–Lithium Exchange

To a solution of 5-bromoindole (3.81 g, 8.0 mmol) in THF (147 mL) at −78 °C was added *t*-BuLi (1.7 M in pentane, 11.4 mL, 19.4 mmol). Following the addition, the reaction mixture was stirred for 15 min at −78 °C and then the boronate (3.6 mL, 17.6 mmol) was added. The mixture was stirred at −78 °C for 1.5 h, allowed to warm to 23 °C, and then quenched with saturated aqueous ammonium chloride (75 mL). The combined organic layers were washed with brine (100 mL), briefly dried over magnesium sulfate, and evaporated under reduced pressure. The crude residue was purified by flash chromatography eluting with hexanes:EtOAc (14:1) to give 3.11 g (74%) of the boronic ester as a yellow oil.

Reference: Garg, N. K.; Sarpong, R.; Stoltz, B. M. *J. Am. Chem. Soc.* **2002,** *124,* 13179–13184.

### 5.1.29.2.2 Standard Suzuki–Miyaura

A solution containing the boronic ester (3.17 g, 6.62 mmol) and the bromide (3.32 g, 13.2 mmol) in benzene (130 mL) and methanol (30 mL), and aqueous sodium carbonate (2 M, 11 mL) was deoxygenated by bubbling a stream of argon through the reaction mixture for 5 min. Tetrakis(triphenylphosphine) palladium(0) (1.15 g, 0.99 mmol) was then added and the flask was equipped with a reflux condenser. The mixture was heated to 80 °C for 2 h and allowed to cool to 23 °C. To the reaction vessel was added sodium sulfate (10 g), which was allowed to stand for 30 min. After filtration over a pad of silica gel ($CH_2Cl_2$ eluent), the filtrate was concentrated to dryness under reduced pressure. The resulting residue was purified by flash chromatography eluting with $CH_2Cl_2$:hexane (1:1) to provide 2.87 g (83%) of the olefin as a yellow oil.

Reference: Garg, N. K.; Sarpong, R.; Stoltz, B. M. *J. Am. Chem. Soc.* **2002,** *124,* 13179–13184.

### 5.1.29.3 Potassium Organotrifluoroborates

Potassium organotrifluorborates are monomeric species that are prepared in one step from their corresponding boronic acids. Also, these compounds exhibit excellent stability in air and may be stored indefinitely. Typically, aryltrifluoroborates couple well to aryl bromides and iodides under ligandless conditions; however, electron deficient coupling partners require the use of Pd(dppf)$Cl_2$. The coupling between aryl- and heteroaryltriflates with trifluoroborates also requires a ligand such as PCy$_3$. Review: Molander, G. A.; Figueroa, R. *Aldrichimica Acta* **2005,** *38,* 49–56.

#### 5.1.29.3.1 Formation of an Organotrifluoroborate

A solution of 3,5-bis(trifluoromethyl)phenylboronic acid (10 g, 38.8 mmol) in methanol (15 mL) was added to an aqueous solution of $KHF_2$ (4.5 M, 26 mL, 117 mmol) at room temperature. A heavy precipitate was deposited. The resulting suspension was stirred for 1 h at room temperature, and the precipitated product was collected and washed with methanol. Recrystallization from a minimum amount of acetone produced 11 g (89%) of potassium 3,5-bis(trifluoromethyl)phenyltrifluoroborate.

Reference: Molander, G. A.; Biolatto, B. *J. Org. Chem.* **2003,** *68*, 4302–4314.

### 5.1.29.3.2 Standard Suzuki

$$\text{Ph-BF}_3\text{K} + \text{1-bromonaphthalene} \xrightarrow[\text{K}_2\text{CO}_3, \text{MeOH}, \Delta]{\text{Pd(OAc)}_2} \text{1-phenylnaphthalene} \quad 75\%$$

To a suspension of potassium phenyltrifluoroborate (92.3 mg, 0.5 mmol), 1-bromonapthalene (103.5 mg, 0.5 mmol), and $K_2CO_3$ (204.5 mg, 1.5 mmol) in MeOH (0.75 mL) was added $Pd(OAc)_2$ in methanol (1.25 mL, $2 \times 10^{-3}$ M) and then the reaction mixture was stirred and heated at reflux for 2 h. The mixture was cooled to room temperature and diluted with water (10 mL). The aqueous phase was extracted with $CH_2Cl_2$ ($3 \times 4$ mL). The combined organic extracts were washed with brine (10 mL) and dried ($MgSO_4$). The crude product was then purified via chromatography (silica gel) eluting with hexane to deliver 76.1 mg (75%) of 1-phenylnaphthalene.

Reference: Molander, G. A.; Biolatto, B. *J. Org. Chem.* **2003,** *68*, 4302–4314.

### 5.1.30 Tsuji–Trost Reaction

Leaving groups in the Tsuji–Trost reaction include acetates, halides, ethers, carbonates, sulfones, carbamates, epoxides, and phosphates. Reviews: (a) Tsuji, J. In *Handbook of Organopalladium Chemistry for Organic Synthesis*; Negishi, E.; deMeijere, A., Eds.; Wiley-Interscience: New York, 2002; Vol II, Palladium-Catalyzed Nucleophilc Substitution Involving Allyl Palladium, Propargyl-palladium and Related Derivatives, pp. 1669–1687. (b) Frost C. G.; Howarth, J.; Williams, J. M. J. *Tetrahedron Asymmetry* **1992,** *3*, 1089–1122.

$$\underset{\text{OMe}}{\overset{\text{OBz}}{\text{pyran}}} \xrightarrow[\text{THF}, \Delta]{\text{NaCH(CO}_2\text{Et)}_2, \text{Pd(PPh}_3)_4} \underset{\text{OMe}}{\overset{\text{CH(CO}_2\text{Et)}_2}{\text{pyran}}} \quad 87\%$$

Sodium hydride (60% oil dispersion, 0.15 g, 0.38 mmol) was washed with hexane ($2 \times 2$ mL) and THF ($1 \times 2$ mL). (*Note: exercise great care with NaH, particularly after the hexane wash; the NaH should be kept under nitrogen or argon.*) To a suspension of oil-free sodium hydride in THF (5mL) was added diethylmalonate (0.770 mL, 0.38 mmol). After 15 min of stirring, the resulting diethyl malonate anion solution was added via a cannula to a solution of the allylic benzoate (131 mg, 0.56 mmol), $Pd(PPh_3)_4$ (45 mg, 0.039 mmol), and triphenylphosphine (100 mg,

0.39 mmol) in THF (5 mL). The resulting orange mixture was heated at 66 °C for 48 h. The reaction mixture was then separated between water (10 mL) and ether (10 mL). The aqueous layer was extracted with ether (3 × 10 mL). The combined organic layers were dried (MgSO$_4$), filtered, and concentrated in vacuo. Purification of the residue by flash chromatography gave 133 mg (87%) of the diester as a colorless oil.

Reference: Brescia, M.-R.; Shimshock, Y. C.; DeShong, P. *J. Org. Chem.* **1997,** *62,* 1257–1263.

## 5.2 Carbon–Carbon Double Bonds (Olefin Formation)

### 5.2.1 Corey–Fuchs Reaction

Treatment of a dibromoolefin with two equivalents of *n*-BuLi will give the corresponding terminal alkyne. Alternatively, the intermediate, lithium acetylide, may be used to trap electrophiles.

#### 5.2.1.1 Formation of a Dibromoalkene

A solution of the aldehyde (10.5 g, 32.2 mmol) in dry CH$_2$Cl$_2$ (400 mL) was treated at room temperature with carbon tetrabromide (42.6 g, 128 mmol) and zinc dust (8.41 g, 128 mmol), followed by triphenylphosphine (33.7g, 128 mmol) in portions to keep the reaction temperature at 25 °C for another 1 h, diluted with hexanes (300 mL), filtered through a thin pad of silica gel, and washed with pentane and ether. The combined organic layers were concentrated in vacuo. The crude product was purified by chromatography (silica gel) eluting with hexanes/EtOAc (10:1) to yield 13.5 g (87%) of the dibromoolefin as a colorless oil.

Reference: Wipf, P.; Xiao, J. *Org. Lett.* **2005,** *7,* 103–106.

#### 5.2.1.2 Formation of a Methylalkyne

A solution of the dibromoolefin (810 mg, 1.68 mmol) in dry THF (20.0 mL) was treated with a solution of *n*-BuLi (1.6 M in hexanes, 2.20 mL, 3.52 mmol). The reaction mixture was stirred at −78 °C for 1 h and at room temperature for another 1 h, cooled to −78 °C, and treated dropwise with methyl iodide (1.05 mL, 16.8 mmol). The resulting mixture was allowed to warm to room temperature and stirred overnight, quenched with water, and extracted with ether. The combined organic layers were dried (MgSO$_4$), concentrated in vacuo, and purified by chromatography on SiO$_2$ eluting with hexanes:EtOAc (15:1) to yield 529 mg (94%) of the methylalkyne as a colorless oil.

Reference: Wipf, P.; Xiao, J. *Org. Lett.* **2005,** *7*, 103–106.

### 5.2.2 Corey–Peterson Olefination

The Corey–Peterson olefination is a version of the Peterson olefination that produces trisubstituted (*E*)-methyl alkenes with high diastereoselectivity via a two-step process.

To a stirred solution of *N*-cyclohexyl-(2-triethylsilylpropylidine)amine (5.13 g, 20.2 mmol) in THF (29 mL) was added *sec*-BuLi (1.4 M in cyclohexane, 13.3 mL, 18.7 mmol). The resulting red–orange mixture was stirred at −78 °C for 30 min and treated with the butyrladehyde (2.52g, 15.6 mmol) in THF (14 mL). The red–orange mixture was immediately warmed to −20 °C and stirred for 1 h and then quenched with water (28 mL). The mixture was then extracted with ether, washed with brine, and dried over MgSO$_4$. The organic layers thus obtained were concentrated and dried under vacuum to give an orange oil (~ 7.35 g). The oil was dissolved in THF (70 mL), cooled to 0 °C, and then treated with trifluoroacetic acid (1.44 mL, 18.7 mmol). The resulting mixture was stirred at 0 °C for 1 h and water (28 mL) was added. This reaction mixture was stirred for 12 h at 0 °C and then carefully quenched with saturated aqueous sodium bicarbonate and extracted with ether. The organic layer was washed with brine, dried of magnesium sulfate, and concentrated. Distillation under reduced pressure gave 2.55 g (81%) of the (*E*)-α,β-unsaturated aldehyde as a clear oil.

Reference: Zeng, X.; Zeng, F.; Negishi, E. *Org. Lett.* **2004,** *6*, 3245–3248.

### 5.2.3 Horner–Wadsworth–Emmons Reaction

The Horner–Wadsworth–Emmons (HWE) reaction involves the addition of a stabilized phosphonate anion to an aldehyde or ketone to afford an intermediate which undergoes an elimination reaction to form predominately the (*E*)-alkene. The HWE reaction has been applied inter- and intramolecularly to simple as well as highly functionalized systems.

Reviews: (a) Kelly, S. E. In *Comprehensive Organic Synthesis*; Trost, B. M.; Fleming, I., Eds.; Pergamon: Oxford, 1991; Vol. 1, Chapter 3.1: Alkene Synthesis, pp. 761–782. (b) Maryanoff. B. E.; Reitz, A. B. *Chem. Rev.* **1989,** *89*, 863–927. (c) Boutagy, J.; Thomas, R. *Chem. Rev.* **1974,** *24*, 87–99.

### 5.2.3.1 Standard Horner–Wadsworth–Emmons

To a solution of trimethylphosphonate (2.35 mL, 14.5 mmol) in THF (26 mL) was added a solution of *n*-BuLi (1.57 M in hexane, 6.94 mL, 10.9 mmol) at 0 °C. The mixture was stirred at room temperature for 15 min, cooled again to 0 °C to which the aldehyde (1.23 g, 3.63 mmol) in THF (10 mL) was added, and the mixture was stirred at 0 °C for 1 h. The reaction was quenched with a pH 7 buffer solution, and the mixture was extracted with EtOAc (6×). The combined organic extracts were washed with brine, dried over magnesium sulfate, and concentrated in vacuo. The crude oil was purified by silica gel chromatography eluting with EtOAc/hexane (40:60 then 50:50) to give 1.32 g (92%) of the ester as a colorless oil.

Reference: Nakamura, R.; Tanino, K.; Miyashita, M. *Org. Lett.* **2003,** *5*, 3579–3582.

### 5.2.3.2 Masamune–Roush Modification of the Horner–Wadsworth–Emmons Reaction

The Masamune–Roush modification of the HWE is a very mild variant that does not require the use of a strong base (i.e., NaH or *n*-BuLi) to generate the phosphonate anion. Instead, in the presence of LiCl, a lithium chelate forms which enhances the acidity of the α-protons. Hence, DBU is sufficiently basic to carry out the deprotonation of the phosphonate.

Since the original paper, milder bases have been used, including triethylamine and disopropylethylamine (Hunigs' base). Again, there is strong preference for the formation of the (*E*)-alkene when the coupling partner is an aldehyde.

To a suspension of LiCl (160 mg, 3.74 mmol) in CH$_3$CN (10 mL) was added trimethyl phosphonoacetate (0.76 mL, 4.68 mmol), and the resulting mixture was stirred for 5 min. Et$_3$N (0.52 mL, 3.74 mmol) was added, followed by 10 min of stirring. A solution of the aldehyde (1.15 g, 3.12 mmol) in CH$_3$CN (5 mL) was added to the reaction mixture which was then stirred overnight. Ether (20 mL) and a saturated solution of ammonium chloride (30 mL) were added, and the layers were separated. The aqueous layer was extracted with ether (3 × 20 mL), and the combined organic extracts were washed with brine, dried over MgSO$_4$, and concentrated. Purification of the crude product was accomplished via chromatography on SiO$_2$ (10% ether/hexane) to give 1.24 g (94%) of the α,β-unsaturated ester as a clear oil.

Reference: Dineen, T. A.; Roush, W. R. *Org. Lett.* **2004,** *6,* 2043–2046.

### 5.2.3.3 Still–Gennari Modification of the Horner–Wadsworth–Emmons Reaction

The Still–Gennari modification of the HWE reaction was a major achievement, because this modification allowed access to (Z)-olefins with high stereoselectivity using a phosphonate. In this version of the HWE, strong preference for Z-alkenes is achieved with the phosphonate bearing electron withdrawing groups under ionic dissociating conditions (crown-ether).

18-Crown-6 (3.51 g, 13.28 mmol) was taken in 30 mL of THF, and the flask was cooled to −78 °C before addition of the phosphonate (2.90, 8.38 mmol) followed by dropwise addition of potassium hexamethyldisilazide (KHMDS 0.5 M in toluene, 16.0 mL, 8.38 mmol). After 10 min, the aldehyde (1.59 g, 6.68 mmol) was added in THF (10 mL). The reaction was quenched after 5 min with NaHCO$_3$ solution. Extractive workup with methyl *tert*-butyl ether followed. The organic layer was then dried over MgSO$_4$ and filtered. The solvent was evaporated in vacuo, and the crude product was chromatographed (hexane/EtOAc, 3:1) to give 1.73 g (85%) of the (Z)-α,β-unsaturated ethyl ester.

Reference: Bhatt, U; Christmann, M.; Quitschalle, M.; Claus, E.; Kalesse, M. *J. Org. Chem.* **2001,** *66,* 1885–1893.

### 5.2.3.4 Ando Modification of the Horner–Wadsworth–Emmons Reaction

The Ando modification of the HWE reaction is yet another major contribution to the formation of (Z)-alkenes via phosphonates. In this version of the HWE, diarylphosphonates are utilized. Review: Ando, K. *J. Org. Chem.* **1998,** *63,* 8411–8416.

To a suspension of NaH (65 mg, 60% dispersion in oil, 2.7 mmol) in anhydrous THF (4.5 mL) was slowly added a solution of the phosphonate (480 mg, 1 mmol) in anhydrous THF (1.5 mL) at −78 °C under argon atmosphere. The reaction mixture was stirred for 15 min at the same temperature. Then, a solution of the aldehyde (492 mg, 1.44 mmol) in THF (1.5 mL) was slowly added, and the reaction mixture was stirred at −78 °C for 40 min. It was brought to −35 °C and quenched with aqueous ammonium chloride. The reaction mixture was extracted with $CH_2Cl_2$. The organic layer was washed with water and brine. It was dried over sodium sulfate and concentrated. The crude residue was chromatographed over silica gel to give 420 mg (73%) of the (Z)-α,β-unsaturated ester as an oil.

Reference: Singh, R. P.; Singh, V. K. *J. Org. Chem.* **2004**, *69*, 3425–3430.

### 5.2.4 Julia Coupling

The Julia coupling sequence is a reliable two-step method for affording (*E*)-disubstituted olefins from a phenyl sulfone, bearing two α-hydrogens, and an aldehyde. The reaction sequence opens with a condensation reaction to furnish an aldol-like product followed by a reductive–elimination process to afford the alkene. The Julia coupling has been applied to the formation of tri- and tetrasubstituted olefins where the coupling partner is a ketone, and a sulfone bears two α-hydrogens or a ketone and sulfone with one α-hydrogen. However, the stereoselectivity is less defined. Reviews: (a) Dumeunier, R.; Markó, I. E. In *Modern Carbonyl Olefination-Methods and Applications*: Takeda, T. Ed.; Wiley-VCH: Weinheim, 2004; Chapter 3: The Julia Reaction, pp. 104–150. (b) Kelly, S. E. In *Comprehensive Organic Synthesis*; Trost, B. M.; Fleming, I., Eds.; Pergamon: Oxford, 1991; Vol. 1, Chapter 3.1: Alkene Synthesis, pp. 743–746. (c) Kocienski, P. *Phosphorus Sulfur* **1985**, *24*, 97–127.

To a solution of the sulfone (4.85 g, 11.98 mmol) in THF (110 mL) at −70 °C was added dropwise *n*-BuLi (1.6 M in hexane, 16.5 mL, 26.36 mmol). After the mixture was stirred for 30 min, a solution of the aldehyde (6.20 g, 15.58 mmol) in THF (10 mL) was added dropwise. The reaction was stirred at the same temperature for 4 h. Saturated ammonium chloride solution was added. The mixture was then extracted with EtOAc multiple times. The combined organic extracts were washed with brine, dried, and evaporated. The crude product was then purified via flash chromatography (silica gel) eluting with heptane/EtOAc (8:1 then 6:1) to afford 7.98 g (83%) of the β-hydroxysulfone as a mixture of two diastereomers.

To a solution of the β-hydroxysulfone diastereomers (13.31 g, 16.58 mmol) in methanol (330 mL) at 0 °C were added $Na_2PO_4$ (28.24 g, 198.9 mmol) and 6% Na–Hg (57.20 g, 149.2 mmol). The reaction mixture was stirred at 0 °C for 2 h. Methanol was evaporated, and the residue was separated in water and EtOAc. The aqueous layer was extracted with EtOAc. (*Note: the aqueous layer must be properly disposed as mercury is present.*) The combined organic layers were washed with brine, dried, and evaporated. The crude product was purified via flash chromatography (silica gel) eluting with heptane:EtOAc (10:1) to afford 7.69 g (72%) of the *E* and *Z*-olefinic isomers as a 78:1 mixture.

Reference: Wang, Q.; Sasaki, A. *J. Org. Chem.* **2004,** *69*, 4767–4773.

### 5.2.5 Knoevenagel Condensation

The Knoevenagel reaction is the condensation of malonic esters or acetoacetic esters (an active methylene) with aldehydes or ketones. The catalysts are mostly organoamines.

Review: Tietze, L. F.; Beifuss, U. In *Comprehensive Organic Synthesis*; Trost, B. M.; Fleming, I., Eds.; Pergamon Press: Oxford, U. K., 1991; Vol. 2, Chapter 1.11: The Knoevenagel Reaction, pp. 341–394.

A mixture of the keto-sulfoxide (4.6 g, 15 mmol), *p*-tolualdehyde (2.0 g, 16.5 mmol), piperidine (0.18 g, 2 mmol), and acetic acid in toluene (40 mL) was heated at reflux for 2 h with azeotropic removal of water using a Dean–Stark trap. The mixture was cooled to room temperature and concentrated in vacuo. The residue was purified by flash chromatography eluting with a hexane:EtOAc:toluene (3:1:1) mixture to give 5.5 g (90%) of the olefin as a solid.

Reference: Swenson, R. E.; Sowin, T. J.; Zhang, H. Q. *J. Org. Chem.* **2002,** *67*, 9182–9185.

### 5.2.6 McMurry Coupling

The McMurray coupling is the reductive coupling between two carbonyl containing compounds, typically aldehydes and ketones, in the presence of $TiCl_3$ or $TiCl_4$. The reaction has been applied inter- and intramolecularly. Reviews: (a) Ephritikhine, M.; Villiers, C. In *Modern Carbonyl Olefination-Methods and Applications*: Takeda, T. Ed.; Wiley-VCH: Weinheim, 2004; Chapter 6: The McMurry Coupling and Related Reactions, pp. 223–285. (b) Ephritikhine, M. *Chem. Commun.* **1998,** *23*, 2549–2554. (c) Fürstner, A; Bogdanović, B. *Angew. Chem. Int. Ed.* **1996,** *35*, 2442–2469. (d) Robertson, G. M. *Comp. Org. Syn.* **1991,** *3*, 583–595; (e) McMurry, J. E. *Chem. Rev.* **1989,** *89*, 1513–1524. (f) McMurry, J. E. *Acc. Chem. Res.* **1983,** *16*, 405–411. (g) Lai, Y.-H. *Org. Prep. Proceed.* **1980,** *12*, 361–391.

$$\text{cyclohexanone} \xrightarrow[\text{DME}]{\text{TiCl}_3(\text{DME})_{1.5},\ \text{Zn-Cu}} \text{cyclohexylidenecyclohexane}$$

97%

*Preparation of TiCl$_3$(DME)$_{1.5}$*

TiCl$_3$ (25.0g, 0.162 mol) was suspended in dry DME (350 mL), and the mixture was refluxed for 2 days under argon. After the mixture was cooled to room temperature, it was filtered under argon, washed with pentane, and dried under vacuum to give 32.0 g (80%) of the fluffy blue crystalline TiCl$_3$(DME)$_{1.5}$.

*Preparation of Zinc–Copper Couple*

Zinc dust (9.8 g, 150 mmol) was added to nitrogen purged water (40 mL). The resulting slurry was then purged with nitrogen for 15 min. CuSO4 (0.75g, 4.7 mmol) was then added. The black slurry was filtered under nitrogen, washed with deoxygenated (nitrogen purged) water, acetone, and ether, and dried under vacuum. The couple can be stored indefinitely in a Schlenk tube under nitrogen.

*Cyclohexylidenecyclohexane*

TiCl$_3$(DME)$_{1.5}$ (5.2 g, 17.9 mmol) and Zn–Cu (4.9g, 69 mmol) were transferred under argon to a flask containing DME (100 mL), and the resulting mixture was refluxed for 2 h to yield a black suspension. Cyclohexanone (0.44 g, 4.5 mmol) in DME (10 mL) was added, and the mixture was refluxed for 8 h. After being cooled to room temperature, the reaction mixture was diluted with pentane (100 mL), filtered through a pad of Florisil, and concentrated to yield 0.36 g (97%) of cyclohexylidenecyclohexane as white crystals. The authors noted that if a 3:1 ratio of TiCl$_3$(DME)$_{1.5}$ to cyclohexanone was used, the yield decreased to 94%. Also, the yield further decreased to 75% when a 2:1 ratio of TiCl$_3$(DME)$_{1.5}$ to cyclohexanone was used.

Reference: McMurry, J. E.; Lectka, T.; Rico, J. G. *J. Org. Chem.* **1989**, *54*, 3748–3749.

## 5.2.7 Methylenation Reagents

### 5.2.7.1 Lombardo–Takai

The Lombardo–Takai olefination is a mild method for converting ketones, particularly enolizable ones, to terminal olefins. Esters are also converted to olefins in the presence of TMEDA. In order for these methylenation conditions to be reproducible, great care must be given during the preparation of the active reagent. The methylenating reagent is formed with zinc powder, a dihalomethane, and titanium chloride. Lombardo reported in the early 1980s that an "ageing" period of 3 days is required prior to use. This simply entails stirring the reagents at 0 °C for 3 days. Reviews: (a) Matsubara, S.; Oshima, K. In *Modern Carbonyl Olefination-Methods and Applications*: Takeda, T., Ed.; Wiley-VCH: Weinheim, 2004; Chapter 5: Olefination of Carbonyl Compounds by Zinc and Chromium Reagents, pp.203–208. (b) Pine, S. H. *Org. React.*, **1993**, *43*, 1–91.

## Reaction Scheme

Substrate (cyclohexanone with OTBS group) → product (methylenated) using Zn, CH$_2$Br$_2$, TiCl$_4$ in THF, CH$_2$Cl$_2$, 91%.

To a suspension of zinc dust (2.87 g, 44 mmol) and CH$_2$Br$_2$ (1.01 mL, 14.4 mmol) in THF (25 mL), stirred under argon atmosphere at –40 °C was added dropwise neat titanium tetrachloride (1.13 mL, 10.3 mmol). The mixture was then allowed to warm to 5 °C and was stirred for 3 days at this temperature to produce a thick gray slurry of the active reagent.

To a solution of the ketone (300 mg, 1.05 mmol) stirred at 5 °C under argon was added the methylenation reagent (2.5 equivalents). After being stirred for 4 h at this temperature, the reaction was diluted with CH$_2$Cl$_2$ and then poured into a cold saturated aqueous solution of sodium bicarbonate. The mixture was extracted with ether. The combined organic extracts were washed with brine, dried, filtered, and concentrated. The residue was subjected to column chromatography to afford 270 mg (91%) of the terminal olefin as an oil.

Reference: Laval, G.; Audran, G.; Galano, J.-M.; Monti, H. *J. Org. Chem.* **2000**, *65*, 3551–3554.

### 5.2.7.2 Petasis Reagent

The Petasis reagent reacts with aldehydes, ketones, and esters to afford terminal olefins. In addition, this reagent enjoys greater air stability than the Tebbe reagent.

*Preparation of dimethyltitanocene*

Cp$_2$TiCl$_2$ → Cp$_2$TiMe$_2$ using MeMgBr, PhMe, 85%.

A 1-L, three-necked, round-bottomed flask, equipped for mechanical stirring, and outfitted with a 250-mL, pressure equalizing addition funnel, a Claisen adapter bearing a thermometer, and a nitrogen inlet/outlet vented through a mineral oil bubbler, was placed under a nitrogen atmosphere and charged with titanocene dichloride (41.5 g, 0.167 mol) and dry toluene (450 mL). The slurry was efficiently stirred and chilled to an internal temperature of –5°C in a ice–methanol bath; 126 mL of a 3 M solution (0.38 mol) of methylmagnesium chloride in THF was then added dropwise via the addition funnel over 1 h at a rate of addition adjusted to maintain an internal temperature below +8 °C. The resulting orange slurry was mechanically stirred at an internal temperature of 0 to +5 °C for 1 h or until the insoluble purple titanocene dichloride was no longer seen in the suspension. The addition funnel was removed and replaced by a rubber septum, and the reaction was assayed by $^1$H NMR. While the reaction was aging at 0° to +5°C, a 2-L, three-necked, round-bottomed flask, equipped for mechanical stirring, and outfitted with a rubber septum, a Claisen adapter bearing a thermometer, and a nitrogen inlet/outlet vented through a mineral oil bubbler, was placed under a nitrogen atmosphere and charged with 117 mL of 6% aqueous ammonium chloride (7.0 g diluted to 117 mL). The solution was chilled to 1–2 °C, with efficient mechanical stirring. When the formation of dimethyltitanocene was judged to be complete, the toluene/THF reaction mixture was quenched into the well-stirred aqueous ammonium chloride solution via a cannula over a period of 1 h, maintaining an internal temperature of 0° to +5°C in both flasks. Toluene (30 mL) was used to rinse the reaction flask. The biphasic mixture was then poured into a 2-L separatory funnel, with another 30 mL of toluene rinse, and the aqueous phase was separated. The organic layer was washed sequentially with 3 portions of cold water (100 mL each) and brine (100 mL) and then dried over anhydrous sodium sulfate ($Na_2SO_4$, 35 g). The organic layer was filtered and carefully concentrated under reduced pressure on a rotary evaporator at a bath temperature of no more than 35°C to a weight of 150 g. The resulting orange solution was assayed by $^1$H NMR to be 20 weight% dimethyltitanocene (29.55 g, 85.0%). If the solution is stored for more than a week, the reagent should be diluted with 160 mL of dry THF, which has a stabilizing effect on the labile reagent. The solution was stored at 0–10°C under nitrogen in a rubber septum-sealed, round-bottomed flask.

A 50-mL, nitrogen-purged, round-bottomed flask was charged with *cis*-ester 1 ((2R-*cis*)-3-(4-fluorophenyl)-4-benzyl-2-morpholinyl 3,5-bis(trifluoromethyl)benzoate) (2.41 g, 4.57 mmol), dimethyltitanocene in toluene (12 mL of a 20% w/w solution in toluene), and titanocene dichloride (71 mg, 0.28 mmol). The red–orange mixture was heated to 80°C and aged in the dark for 5.5 h, and then cooled to ambient temperature. Sodium bicarbonate (0.60 g), methanol (9.6 mL), and water (0.36 mL) were added, and the mixture was heated to 40°C for 14h. (The hot aqueous methanol treatment was done to decompose the titanium residues into an insoluble solid. The decomposition was judged to be complete when gas evolution ceased.) The green mixture was cooled to ambient temperature and the titanium residues were removed by filtration. The solution was evaporated under reduced pressure and flushed with methanol. The crude material was recrystallized by dissolving in hot (60 °C) methanol (24 mL), cooling to ambient temperature, and then adding water (7.2 mL) over 2 h. The material was stirred for 18 h and then isolated via filtration at ambient temperature. The filter cake was washed with 25% aqueous methanol (6 mL), and the solid was dried at ambient temperature under nitrogen. Vinyl ether 2 ((2R-cis)-2-[[1-[3,5-bis(trifluoromethyl)phenyl]ethenyl]oxy]-3-(4-fluorophenyl)-4-benzylmorpholine) (2.31 g, 96%) was isolated as a pale yellow solid.

Reference: Payack, J. F.; Hughes, D. L.; Cai, D.; Cottrell, I. F.; Verhoeven, T. R. *Org. Syn.* **2004,** *Coll. Vol. 10*, 355–357.

## 5.2.7.3 Tebbe Reagent

The Tebbe reagent offers distinct advantages over the traditional Wittig reaction. First, aldehydes (to a limited extent), ketones, and esters can be converted to terminal olefins. In addition, the low basicity of this commercially available titanocene–methylidine reagent allows for smooth conversion of enolizable ketones to olefins without compromising an α-stereogenic center. Reviews: (a) Takeda, T.; Tsubouchi, A. In *Modern Carbonyl Olefination-Methods and Applications*: Takeda, T. Ed.; Wiley-VCH: Weinheim, 2004; Chapter 4: Carbonyl Olefination Utilizing Metal Carbene Complexes, pp.151–199. (b) Pines, S. H. *Org. React.* **1993**, *43*, 1–91. (c) Kelly, S. E. *Comp. Org. Syn.* **1991**, *1*, 743–746. (d) Kociensky, P. J. *Phos. Sulfur* **1985**, *24*, 97. (e) Brown-Wensley, K. A.; Buchwald, S. L.; Cannizzo, L.; Clawson, L.; Ho, S.; Meinhardt, D.; Stille, J. R.; Straus, D.; Grubbs, R. H. *Pure Appl. Chem.* **1983**, *55*, 1733–1744.

To a solution of the aldehyde (2.56 g, 4.40 mmol) in THF (50 mL) cooled to 0 °C was added the Tebbe reagent (0.5 M, 13.2 mL, 6.60 mmol). After being stirred at 0 °C for 30 min, the reaction mixture was treated with 0.1 M aqueous NaOH (30 mL) and diluted with ether. The resulting mixture was stirred at room temperature for 30 min, filtered through a pad of Celite, and the filtrate was concentrated under reduced pressure. The residue was purified by column chromatography (silica gel, 10% EtOAc/hexanes) to give 2.30 g (90%) of the terminal olefin as a colorless oil.

Reference: Fuwa, H.; Kainuma, N.; Tachibana, K.; Sasaki, M. *J. Am. Chem. Soc.* **2002**, *124*, 14983–14992.

## 5.2.7.4 Wittig

The unstabilized ylide, derived from triphenylphosphonium methyl bromide, is a reagent for introducing terminal olefins. Homologues of the methyl derivative deliver Z-alkenes. See section 5.2.11 for the Wittig reaction.

To a suspension of methyltriphenylphosponium bromide (7.84 g, 21.6 mmol) in dry THF (10 mL) at −78 °C under argon was added *n*-BuLi (10.45 mL of 1.98 M solution in hexanes, 20.7 mmol). The mixture was warmed to 0 °C and stirred for 1 h. The mixture was then re-cooled to −78 °C, and a solution of the ketone (6.30 g, 19.77 mmol) in THF (90 mL) was added slowly. The resulting reaction mixture was allowed to warm to 0 °C and stirred for 5 h. The reaction was quenched by addition of aqueous ammonium chloride, and the mixture was diluted with ether. (*Note: the quenching should be done at 0 °C.*) The phases were separated, and the aqueous phase extracted with ether. The organic phases were washed with brine, dried over sodium sulfate, and concentrated in vacuo. The resulting residue was purified via flash chromatography (hexane:EtOAc, 10:1) to afford 5.25 g (84%) of the diene.

Reference: White, J. D.; Wang, G.; Quaranta, L. *Org. Lett.* **2003**, *5*, 4983–4986.

### 5.2.8 Olefin Metathesis

Although olefin metathesis found its genesis in the 1960s, it was not until the 1990s with the advent of Schrock's molydbedum and Grubbs ruthenium catalysts that the utility of this transformation become apparent.

Schrock's catalyst

1st Generation Grubbs' catalyst   2nd Generation Grubbs' catalyst   Hoveyda-Grubbs' catalyst

In general, olefin metathesis is the exchange of alkylidine groups of alkenes. The four major types of olefin metathesis include ring-closing metathesis (RCM), cross-metathesis (CM), ring-opening metathesis (ROM), and acyclic diene metathesis polymerization (ADMET). For this section, RCM will be the focus along with a procedure for CM using the Hoveyda–Grubbs catalyst. The Schrock and Grubbs catalysts have been extremely successful for RCMs in natural product synthesis for the formation of five, six, seven, eight, nine, ten-membered and higher ring systems. Although both the Mo and Ru catalysts have enjoyed success, the Grubbs catalysts have seen more extensive use because they are easier to handle in the laboratory. They are stable to moisture and air. In contrast, the Schrock catalyst requires manipulation in a glove box. Moreover, the Grubbs' catalysts tolerate a wider range of functional groups in complex oxygenated and alkaloidal molecules. Functional groups include ketones, esters, amides, epoxides, acetals, unprotected alcohols, silyl ethers, amines, and sulfides. In addition to the cyclization dienes to form mono-olefins, RCM has been utilized for enynes to deliver dienes. Reviews: (a) Brenneman, J. B.; Martin, S. F. *Curr. Org. Chem.* **2005,** *9*, 1535–1549. (b) Dieters, A.; Martin, S. F. *Chem Rev.* **2004,** *104*, 2199–2238; (c) McReynolds, M. D.; Dougherty, J. M.; Hanson, P. R. *Chem Rev.* **2004,** *104*, 2239–2258. (d) Grubbs, R. H. *Tetrahedron* **2004,** *60*, 7117–7140. (e) In *Handbook of Metathesis*; Grubbs, R., Ed.;Wiley-VCH: Weinheim, 2003; Vol. 2, pp.1–510. (f) Walters, M. A. Recent Advances in the Synthesis of Heterocycles via RCM. In *Progress in Heterocyclic Chemistry*; Gribble, G. W.; Jouele, J. A., Eds.; Pergamon: Elmsford, NY, 2003; Vol. 15, pp. 1–36. (g) Connon, S. J.; Blechert, S. *Angew. Chem. Int. Ed.* **2003,** *42*, 1900–1923. (h) Trnka, T. M.; Grubbs, R. H. *Acc. Chem. Res.* **2001,** *34*, 18–29. (i) Fürstner, A. *Angew. Chem. Int. Ed.* **2000,** *39*, 3012. (j) Wright, D. L. *Curr. Org. Chem.* **1999,** *3*, 211–240. (k) In *Alkene Metathesis in Organic Synthesis*.; Fürstner, A., Ed. Springer: Berlin, 1998. (l) Grubbs, R. H.; Chang, S. *Tetrahedron* **1998,** *54*, 4413–4450. (m) Boger, D. L.; Chai, W. *Tetrahedron* **1998,** *54*,, 3955–3970. (n) Grubbs, R. H.; Miller, S. J.; Fu, G. C. *Acc. Chem. Res.* **1995,** *28*, 446–452.

*5.2.8.1 Schrock's Catalyst for Ring Closing Metathesis*

To a degassed solution of the diene (130 mg, 0.34 mmol) in benzene (4.0 mL) under nitrogen atmosphere at 20 °C was added Schrock's molybdenum catalyst in benzene (0.13 mL of 100 mg of catalyst in 1.0 mL of benzene, 0.02 mmol). (*Note: it is recommended that this catalyst is handled in a glove box or glove bag under an inert atmosphere*.) The solution was heated to 60 °C for 1 h, at which point TLC analysis indicated that the reaction was complete. The reaction mixture was concentrated. The crude product was purified via chromatography eluting with 20% ether in hexanes to afford 105 mg (88%) of the *bis*-olefin as a colorless oil.

Reference: Kozmin, S. A.; Iwama, T.; Huang, Y.; Rawal, V. H. *J. Am. Chem. Soc.* **2002,** *124*, 4628–4641.

*5.2.8.2 Grubbs' Catalysts for Ring Closing Metathesis*

*5.2.8.2.1 First Generation Grubbs'catalyst*

To a solution of the diene (2.70 g, 11.4 mmol) in $CH_2Cl_2$ (300 mL) at reflux under argon was added bis(tricyclohexylphosphine)benzylidene ruthenium (IV) dichloride

(468 mg, 0.57 mmol), portionwise over a period of 4 h. The mixture was stirred under an air atmosphere for 5 h before being filtered through a 10-cm pad of Florisil/silica. The pad was washed with $CH_2Cl_2$ (2 × 50 mL), and the solvent was removed by evaporation. The residue was purified by column chromatography (EtOAc:hexanes, 90:10) to yield 2.14 g (96%) of the olefin as transparent colorless needles.

Reference: Hanessian, S.; Sailes, H.; Munro, A.; Therrien, E. *J. Org. Chem.* **2003**, *68*, 7219–7233.

### 5.2.8.2.2a  Second Generation Grubbs' Catalyst: Monoalkene Formation from a Diene

| Catalyst | eq. of catalyst | Time | Yield |
|---|---|---|---|
| 1st generation | 20 mol% (5X4 mol%) | 52 h | 46% |
| 2nd generation | 5 mol% | 2 h | 92% |

To a solution of the diene (220 mg, 0.62 mmol) in $CH_2Cl_2$ (16 mL) was added the second generation ruthenium catalyst (28 mg, 0.032 mmol). The reaction was heated at reflux for 2 h. After cooling to room temperature, the reaction mixture was filtered through a pad of $SiO_2$ eluting with EtOAc. The filtrate was concentrated in vacuo, and the crude product was purified on $SiO_2$ eluting with hexanes:EtOAc (4:1 to 2:1) to give 140 mg (92%) of the olefin as a tan solid.

Reference: Wipf, P.; Spencer, S. R. *J. Am. Chem. Soc.* **2005**, *127*, 225–235.

### 5.2.8.2.2b  Second Generation Grubbs' Catalyst: Diene Formation from an Enyne

To a degassed solution of Grubbs II (212 mg, 0.250 mmol) in methylene chloride (500 mL) was added a solution of enyne (744 mg, 2.50 mmol) in degassed methylene chloride (30 mL). The mixture was stirred under a blanket of argon for 16 h, and then DMSO (0.89 mL) was added to decompose the catalyst. The mixture was stirred for an additional 23 h, whereupon the solvent was removed under reduced pressure, and the resulting residue was purified by flash chromatography ($SiO_2$) eluting with ether/pentane (1:2) to afford 623 mg (84%) of the diene as a yellow oil.

Reference: Brenneman, J. B.; Machauer, R.; Martin, S. F. *Tetrahedron* **2004**, *60*, 7301–7314.

### 5.2.8.3 Hoyveda–Grubbs' Catalyst (Intermolecular Reaction)

[Scheme: Homoallylic alcohol with OH and OTBDPS groups + methyl acrylate (MeO-C(O)-CH=CH₂), Hoveda-Grubbs catalyst, CH₂Cl₂, Δ, 86% → enoate product MeO-C(O)-CH=CH-CH₂-CH(OH)-CH₂-CH₂-OTBDPS]

To a stirred solution of the homoallylic alcohol (14.5 g, 40.9 mmol) in CH$_2$Cl$_2$ (100 mL) was added methyl acrylate (11.0 mL, 123 mmol) followed by the catalyst (0.26 g, 0.41 mmol). The resulting mixture was heated to reflux for 5 h, after which another portion of the catalyst was added, and the mixture was heated to reflux for an additional 2 h. The solvent was evaporated and the crude product purified by chromatography on SiO$_2$ (EtOAc:hexanes, 1:3) to give 14.6 g (86%) of the enone as a slightly yellowish oil.

Reference: Dineen, T. A.; Roush, W. R. *Org. Lett.* **2004**, *6*, 2043–2046.

### 5.2.9 Peterson Olefination

The Peterson olefination is a two-step process for the formation of alkenes from an α-silylcarbanion and an aldehyde or ketone. The first step is an addition reaction that affords both *syn* and *anti* β-hydroxysilanes. The stereochemistry is then controlled during the elimination step by using either an acid or a base. Reviews: Kano, N.; Kawashima, T. In *Modern Carbonyl Olefination-Methods and Applications*: Takeda, T. Ed.; Wiley-VCH: Weinheim, 2004; Chapter 2: The Peterson and Related Reactions, pp. 18–103. (b) Kelly, S. E. In *Comprehensive Organic Synthesis*; Trost, B. M.; Fleming, I., Eds.; Pergamon: Oxford, 1991; Vol. 1, Chapter 3.1: Alkene Synthesis, pp. 731–737. (c) Ager, D. J. *Org. React.* **1990**, *38*, 1–223. (d) Ager, D. J. *Synthesis* **1984**, 384–398.

[Scheme: Aldehyde with dioxolane-protected cyclohexanone + TMS-CH(Me)-C(O)-OEt, LDA, THF, −78 °C, 78% (E:Z, 1:1) → alkene product with R₁ = CO₂Et, R₂ = Me and R₂ = Me, R₁ = CO₂Et]

R$_1$ = CO$_2$Et, R$_2$ = Me
R$_2$ = Me, R$_1$ = CO$_2$Et

*Lithium Diisopropylamide formation*

To a solution of diisopropylamine (1.60 mL, 11.4 mmol) in THF (15 mL) at 0 °C was added *n*-BuLi (1.0 M in hexane, 13 mL, 13.0 mmol). The solution was stirred for 15 min at 0 °C and then cooled to −78 °C. Ethyl 2-(trimethylsilyl)propionate (2.00 g,

11.5 mmol) in THF (20 mL) was added dropwise via an addition funnel, and the mixture was stirred for 1 h. The aldehyde (1.97 g, 9.27 mmol) in THF (12 mL) was added dropwise via an addition funnel over a 35-min period, and the mixture was stirred at −78 °C for 3 h, followed by an additional 18 h during which the temperature was slowly raised to room temperature. The reaction was quenched with saturated $NH_4Cl$ (12 mL), and the solution was diluted with ether (150 mL). The organic phase was washed with brine (2 × 15 mL), and the combined aqueous washes were extracted with ether (2 × 10 mL). The combined organic phases were dried ($Na_2SO_4$) and concentrated in vacuo. The residue was chromatographed over 50 g of silica gel eluting with petroleum ether:EtOAc 99:1 to afford 2.14 g (78%) of (*E*)- and (*Z*)-olefins as a 1:1 mixture. (*Note: the authors reported an 8:1 mixture of E:Z when the corresponding stabilized ylide was used*).

Reference: Guevel, A.-C.; Hart, D. J. *J. Org. Chem.* **1996,** *61*, 465–472.

### 5.2.10 Takai Olefination

Review: Matsubara, S.; Oshima, K. In *Modern Carbonyl Olefination-Methods and Applications*: Takeda, T., Ed.; Wiley-VCH: Weinheim, 2004; Chapter 5: Olefination of Carbonyl Compounds by Zinc and Chromium Reagents, pp. 214–220.

$CrCl_2, CHI_3$
dioxane, THF
74%
(*E:Z*, 7.7.:1)

*E* : $R_1$ = I, $R_2$ = H
*Z*:: $R_1$ = H, $R_2$ = I

To a slurry of flame-dried chromium (II) chloride (10.42 g, 84.8 mmol) in THF (20 mL) and dioxane (60 mL) was added the aldehyde (4.71 g, 11.3 mmol) and iodoform (8.90 g, 22.6 mmol) in dioxane (60 mL). The resulting brown suspension was stirred at room temperature for 10 h, diluted with ether (300 mL), and poured into water (300 mL). The aqueous layer was separated and washed with saturated sodium chloride. The aqueous layer thus afforded was extracted with ether (500 mL). The combined organic extracts were washed with brine (200 mL), dried over magnesium sulfate, filtered, and concentrated. Flash chromatography (hexanes: $CH_2Cl_2$, 7:1 to 6:1) provided 4.0 g (65.5%) of the *E*-olefin and 0.52 g (8.5%) of the Z-olefin.

Reference: Song, H. Y.; Joo, J. M.; Kang, J. W.; Kim, D.-S.; Jung, C.-K.; Kwak, H. S.; Park, J. H.; Lee, E.; Hong, C. Y.; Jeong, S.; Jeon, K.; Park, J. H. *J. Org. Chem.* **2003,** *68*, 8080–8087.

### 5.2.11 Wittig Olefination

The Witttig olefination and its variants have been one of the most indispensable reactions to the synthetic organic chemist in the past 50 years for the generation of olefins from aldehydes and ketones. In addition, the stereochemistry of the olefin is often well predicted, and a chemical handle is often introduced which provides for homologation or functional group manipulation. The basic Wittig reaction is between an ylide ($Ph_3P=CR_1R_2 \leftrightarrow PPh_3^+-{}^-CR_1R_2$) and an aldehyde or ketone. Ylides are relatively stable species which may be generated from the corresponding triphenylphosphonium salts and a base (i.e., *n*-BuLi, NaH, LiHMDS, *etc*.). In general, unstabilized ylides deliver the (*Z*)-alkene while stabilized ylides give the (*E*)-alkene. Examples of unstabilized and stabilized ylides are given in the subsequent procedures. Reviews: (a) Edmonds, M.; Abell, A. In *Modern Carbonyl Olefination-Methods and Applications*: Takeda, T. Ed.; Wiley-VCH: Weinheim, 2004; Chapter 1: The Wittig Reaction, pp. 1–17. (b) Kolodiazhny, O. I. In *Phosphorous Ylides Chemistry and Application in Organic Synthesis*; Wiley-VCH: Weinheim, 1999; pp. 360–538. (c) Maryanoff, B. E.; Reitz, A. B. *Chem. Rev.* **1989**, *89*, 863–927. (d) Bestmann, H. J.; Vostrowsky, O. *Top. Curr. Chem.* **1983**, *109*, 85–163. (e) Gosney, I.; Rowley, A. G. In *Organophosphorus Reagents in Organic Synthesis*, Cadogan J. I. G., Ed.; Academic Press: New York, 1979; pp 17–153; (f) Wadsworth, W. S., Jr., *Org. React.* **1977**, *25*, 73–253. (g) Vollhardt, K. P. C. *Synthesis* **1975**, 765–780. (g) Maercker, A. *Org. React.* **1965**, *14*, 270–490.

### 5.2.11.1 Unstabilized Ylides

Unstabilized ylides deliver the (*Z*)-alkene.

PhS–CH(CH$_3$)–CHO  →  [EtPPh$_3$I, *n*-BuLi, THF, −78 °C to rt, 85%]  →  PhS–CH(CH$_3$)–CH=CH–CH$_3$ (*cis*)

To a suspension of ethylphosphonium iodide (4.1 g, 11 mmol) in dry THF (40 mL) at 0 °C was added a solution of *n*-BuLi (1 M, 11 mL, 11 mmol). The mixture was stirred for 30 min at 0 °C. It was then cooled to −78 °C, and a solution of the aldehyde (1.8 g, 10 mmol) in dry THF (22 mL) was added dropwise. The reaction mixture was stirred for 4 h while allowing to gradually reach room temperature. It was cooled to 0 °C and quenched by the addition of saturated aqueous ammonium chloride. The layers were separated, and the aqueous layer extracted with ether. The combined organic layers were washed successively with water and brine, and dried over sodium sulfate. The organic extracts were concentrated, and the resulting crude product was purified by column chromatography eluting with petroleum ether/EtOAc (95:5) to afford 1.63 g (85%) of the *cis*-olefin as a liquid.

Reference: Raghavan, S.; Reddy, S. R.; Tony, K. A.; Kumar, C. N.; Varma, A. K., Nangia, A. *J. Org. Chem.* **2002**, *67*, 5838–5841.

### 5.2.11.2 Stabilized Ylides

Stabilized ylides give predominately the (*E*)-olefin.

TBSO/TES-substituted aldehyde  →  [Ph$_3$P=C(CH$_3$)CO$_2$Et, PhMe, 100 °C, 99% (*E:Z* >9:1)]  →  TBSO/TES-substituted α,β-unsaturated ester (CO$_2$Et)

The acetylenic aldehyde (3.4 g, 10.0 mmol) was dissolved in toluene (200 mL). (Carbethoxymethylene)triphenylphosphorane (18.1 g, 50.0 mmol) was added in 1 portion. The stirred mixture was warmed to 100 °C and kept at this temperature for 12 h. The homogeneous solution was cooled to room temperature and washed sequentially with water (50 mL) and brine (50 mL). The organic phase was dried over sodium sulfate, filtered, concentrated, and the crude mixture purified by chromatography (silica gel, hexanes:EtOAc, 98:2) to afford 4.2 g (99%, $E{:}Z > 9{:}1$) of the $E$-olefin.

Reference: Nicolaou, K. C.; Fylaktakidou, K. C.; Monenschein, H.; Li, Y.; Weyershausen, B.; Mitchell, H. J.; Wei, H.-X. Guntupalli, P.; Hepworth, G.; Sugita, K. *J. Am. Chem. Soc.* **2003**, *125*, 15433–15442.

## 5.3 Reactions that Form Multiple Carbon–Carbon Bonds

### 5.3.1 Diels–Alder Reaction Using Danishefsky's Diene

The Danishefsky's diene reacts readily with dienophiles, including alkenes, alkynes, aldehydes, and imines. Reviews: (a) Oppolzer, W. In *Comprehensive Organic Synthesis;* Trost, B. M.; Fleming, I., Eds.; Pergamon: Oxford, 1991; Vol. 5, Chapter 4.1: Intermolecular Diels-Alder Reactions, pp. 315–399. (b) Boger, D. L. In *Comprehensive Organic Synthesis;* Trost, B. M.; Fleming, I., Eds.; Pergamon: Oxford, 1991; Vol. 5, Chapter 4.3: Heterodiene Additions, pp. 451–512. (c) Weinreb, S. M. In *Comprehensive Organic Synthesis;* Trost, B. M.; Fleming, I., Eds.; Pergamon: Oxford, 1991; Vol. 5, Chapter 4.2: Heterodienophile additions to Dienes, pp. 401–499. (d) Mehta, G.; Uma, R. *Acc. Chem. Res.* **2000,** *33*, 278. (e) Jorgensen, K. A. *Eur. J. Org. Chem.* **2002**, 2093. (f) Corey, E. J. *Angew. Chem., Int. Ed.* **2002**, *41*, 1650.

To a cooled solution of the enone (2 mmol) and Danishefsky's diene (480 mg, 2.6 mmol) in toluene (18 mL) was added Yb(OTf)$_3$ (31 mg, 0.05 mmol) at 0 °C. The mixture was stirred at the same temperature for 24 h. The reaction mixture was washed with 2 M K$_2$CO$_3$ and dried (MgSO$_4$). After filtration and concentration of the filtrate, the residue was purified via flash chromatography using silica gel eluting with EtOAc/hexanes (20–30%) to give 428 mg (91%) of the adduct.

Reference: Inokuchi, T.; Okano, M.; Miyamoto, T.; Madon, H. B.; Takagi, M. *Synlett* **2000,** 1549–1552.

### 5.3.2 Hetero-Diels-Alder Reaction Using Rawal's Diene

To a 25-mL flame-dried flask under a nitrogen atmosphere were added freshly distilled Rawal's diene (227 mg, 1 mmol, 1.0 equivalents) and 2 mL of CHCl$_3$. Benzaldehyde (1.5 mmol, 1.5 equivalents) was added dropwise via a gas-tight syringe. The reaction mixture was stirred at room temperature for 2 h and then diluted with 15 mL of CH$_2$Cl$_2$. The yellow solution was cooled to −78 °C and treated with 142 µL acetyl chloride (2 mmol, 2 equivalents). After stirring for about 30 min, saturated Na$_2$CO$_3$ was added. The organic layer was separated and the water phase was diluted with 15 mL of water and extracted twice with CH$_2$Cl$_2$. The combined organic phase was dried with MgSO$_4$, filtered, and concentrated to give a yellow oil, which was subjected to flash chromatography to afford the desired dihydropyrone in 86% overall yield.

Reference: Huang, Y.; Rawal, V. H. *Org. Lett.* **2000,** *2,* 3321–3322.

### 5.3.3 Simmons–Smith Reaction

The Simmons–Smith cyclopropanation converts olefins to the corresponding cyclopropanes stereospecifically. Reviews: (a) Charette, A. B.; Beauchemin, A. *Org. React.* **2001,** *58,* 1–415. (b) For enantioselective cyclopropanations see Davies, H. M. L.; Antoulinakis, E. G. *Org. React.* **2001,** *57,* 1–326.

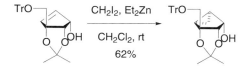

To a stirred solution of the cyclopentenol (2.14g, 4.99 mmol) in CH$_2$Cl$_2$ (40 mL) was added diethylzinc (1.0 M in hexane, 25 mL, 25.00 mL) at 0 °C, and the reaction mixture was stirred at the same temperature for 15 min. Diiodomethane (4.0 mL, 50.21 mmol) was added to the reaction mixture at 0 °C, and the resulting mixture was stirred at room temperature overnight. After aqueous ammonium chloride solution (10 mL) was added, the reaction mixture was partitioned between ethyl acetate and water. The layers were separated, and the organic layer was dried (MgSO$_4$), filtered, and evaporated in vacuo. The residue was purified by silica gel column chromatography using hexane and ethyl acetate (2.5:1) as the eluent to give 1.38 g (62%) of the cyclopropane as a white solid.

Reference: Lee, J. A.; Moon, H. R.; Kim, H. O.; Kim, K, R.; Lee, K. M.; Kim, B. T.; Hwang, K. J.; Chun, M. W.; Jacobson, K. A.; Jeong, L. S. *J. Org. Chem.* **2005,** *70,* 5006–5013.

# 6

# Protecting Groups

In his book, *Protecting Groups*, Philip J. Kocieński stated that there are three things that cannot be avoided: *death, taxes, and protecting groups*. Indeed, protecting groups mask functionality that would otherwise be compromised or interfere with a given reaction, making them a necessity in organic synthesis. In this chapter, for each protecting group showcased, only the most widely used methods for protection and cleavage are shown. Also, this section is not comprehensive and only addresses some of the most common blocking groups in organic synthesis. For a thorough review of protecting groups, the reader should consult the following references: (a) Wuts, P. G. M.; Greene, T. W.; *Protective Groups in Organic Synthesis*, 4th ed.; Wiley: Hoboken, NJ, 2007; (b) Kocienski, P. J. *Protecting Groups*, 3rd edition.; Thieme: Stuggart, 2004.

## 6.1 Alcohols and Phenols

In this section, the formation and cleavage of eight protecting groups for alcohols and phenols are presented: acetate; acetonides for diols; benzyl ether; *para*-methoxybenzyl (PMB) ether; methyl ether; methoxymethylene (MOM) ether; *tert*-butyldiphenylsilyl (TBDPS) silyl ether; and tetrahydropyran (THP).

### 6.1.1 Acetate

Acetate is a convenient protecting group for alcohols—easy on and easy off. Selective protection of a primary alcohol in the presence of a secondary alcohol can be achieved at low temperature. The drawback of this protecting group is its incompatibility with hydrolysis and reductive conditions.

*Protection*

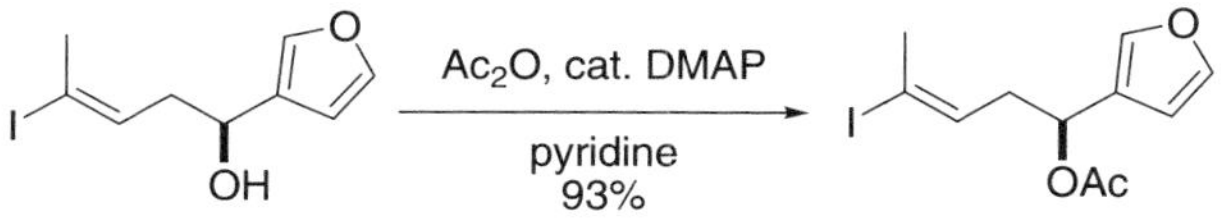

A solution of (+)-(*E*)-1-(3-furyl)-4-iodopent-3-en-1-ol (58 mg, 0.21 mmol), acetic anhydride (0.04 mL, 0.421 mmol), 4-(dimethylamino)pyridine (DMAP, 1.2 mg, 0.01 mmol), and pyridine (1 mL) was stirred until disappearance of the starting material. The mixture was quenched with a saturated aqueous $NaHCO_3$ solution. The aqueous layer was extracted with diethyl ether and the organic layers were washed with a saturated aqueous $CuSO_4$ solution and water, dried over $MgSO_4$, and concentrated under vacuum. The crude product was purified by flash chromatography (light petroleum/$Et_2O$, 9:1) to give (+)-(*E*)-1-(3-furyl)-4-iodopent-3-en-1-yl acetate at 93% yield.

Reference: Commeiras, L.; Parrain, J.-L. *Tetrahedron: Asymmetry* **2004**, *15*, 509–517.

*Cleavage*

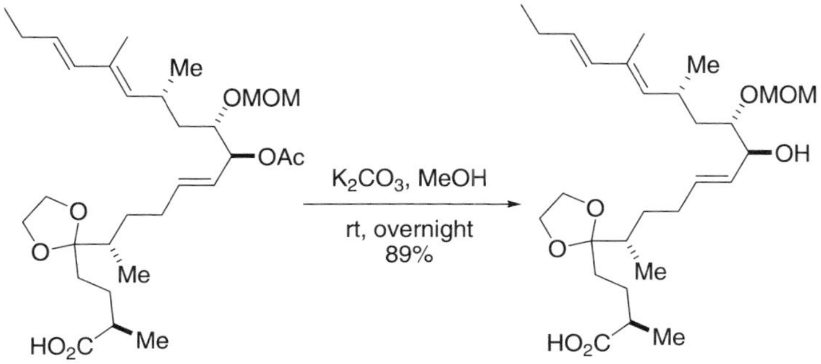

To a solution of the acetate starting material (78 mg, 0.145 mmol) in MeOH (4 mL) was added $K_2CO_3$ (400 mg, 2.9 mmol). The reaction was stirred at room temperature overnight. The mixture was poured into pH 4 buffer (aqueous solution of $NaH_2PO_4$ and $NaHSO_4$) and diluted with $CH_2Cl_2$. The aqueous layer was extracted with $CH_2Cl_2$ three times. The combined organic layers were dried over $MgSO_4$ and concentrated in vacuo. Purification by flash chromatography (40% EtOAc/hexanes) afforded the corresponding alcohol (64 mg, 89%) as a clear oil.

Reference: Ghosh, A. K.; Gong, G. *J. Am. Chem. Soc.* **2004**, *126*, 3704–3705.

### 6.1.2 Acetonide

*O*,*O*-Isopropylidene acetal has been widely used in protecting 1,2- and 1,3-diols. They are resistant to very basic conditions, but are cleaved under acidic conditions.

*Protection*

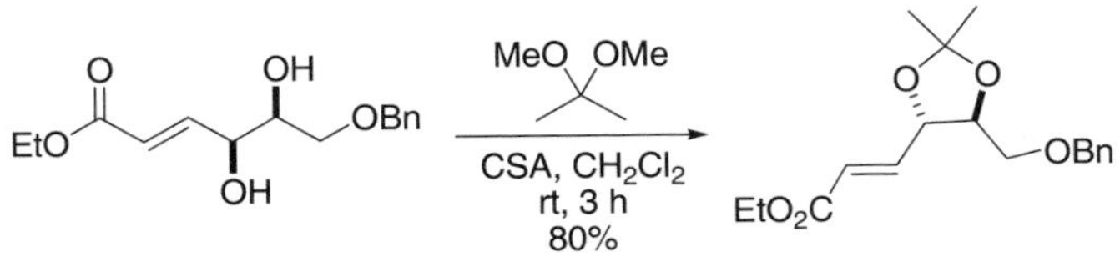

To a stirred solution of the diol starting material (0.8 g, 2.85 mmol) in CH$_2$Cl$_2$ (2 mL) at room temperature was added 2,2,-dimethoxypropane (0.52 mL, 4.2 mmol) and camphorsulfonic acid (CSA, 13 mg, 2 mol%). The reaction was stirred for 3 h and quenched with saturated aqueous sodium bicarbonate (10 mL) and the aqueous layer was extracted with ether (3 × 15 mL). The combined organic layers were washed with brine (10 mL), dried (Na$_2$SO$_4$), filtered, and concentrated in vacuo. Purification of the residue by flash chromatography eluting with EtOAc/hexane (1:9) afforded the isopropylidene acetal (0.73 g, 80%) as a viscous oil.

Reference: Ahmed, M. M.; Berry, B. P.; Hunter, T. J.; Tomcik, D. J.; O'Doherty, G. A. *Org. Lett.* **2005**, *7*, 745–748.

Protection of the 1,3-diol as an acetonide works similarly to the 1,2-diol.

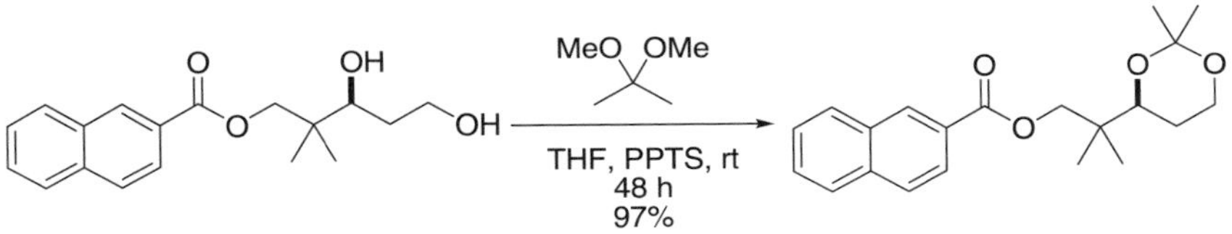

A solution of this ester (8.35 g, 27.6 mmol, 1.0 equivalents) in tetrahydrofuran (THF 100 mL) was cooled to 0 °C, and pyridinium *p*-toluenesulfonate (PPTS, 500 mg, 2.00 mmol, 0.1 equivalents) and then 2,2-dimethoxypropane (20.0 mL, 163 mmol, 5.9 equivalents) were added. The cold bath was removed and the mixture was stirred at room temperature for 48 h, quenched with saturated aqueous NaHCO$_3$ solution, and extracted with EtOAc. The combined organic layers were washed with brine, dried (Na$_2$SO$_4$), filtered, and concentrated. Purification by chromatography on SiO$_2$ (3% ethyl acetate and 1% triethylamine in hexanes, and then 100% EtOAc) afforded the acetonide and a small amount of starting diol which was re-subjected and purified as above. The two batches were combined to afford naphthalene-2-carboxylic acid 2-[(4*S*)-2,2- dimethyl-[1,3]dioxan-4-yl]-2-methylpropyl ester (9.140 g, 97%) as a clear colorless syrup.
Reference: Wipf, P.; Graham, T. H. *J. Am. Chem. Soc.* **2004**, *126*, 15346–15347.

*Cleavage*

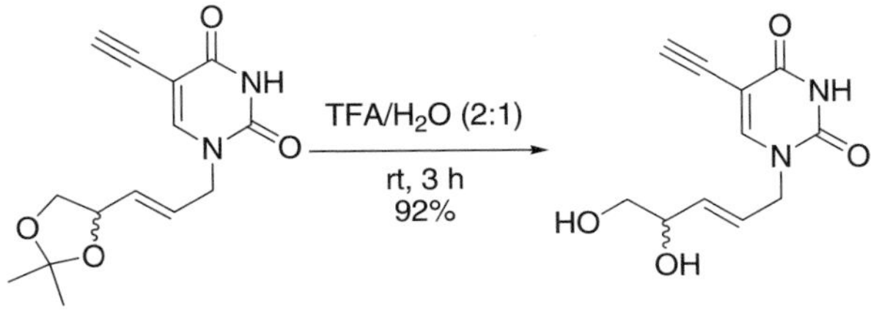

The acetal derivative (0.22 mmol) was stirred at room temperature for 3 h in a 2:1 mixture of trifluoroacetic acid (TFA)/H$_2$O (15 mL). After evaporation of volatiles, the crude residue was purified by flash chromatography (CH$_2$Cl$_2$/MeOH, 92:8) to give the corresponding diol (92% yield) as a pale yellow gum.

Reference: Amblard, F.; Aucagne, V.; Guenot, P.; Schinazi, R. F.; Agrofoglio, L. *Bioorg. Med. Chem.* **2005**, *13*, 1239–1248.

### 6.1.3 Benzyl ether

Benzyl chloride and benzyl bromide are strong lachrymators and therefore their usage should be carried out in the hood. When benzyl chloride is used in making the benzyl ether, addition of a catalytic amount of KI speeds up the reaction (Finkelstein reaction).

*Protection*

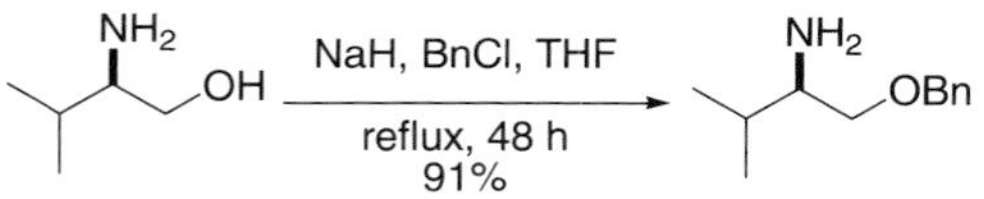

To a stirred solution of (*R*)-valinol (5.0 g, 48.5 mmol) in THF (50 mL) was added sodium hydride (60% dispersion, 1.94 g, 48.5 mmol) in 1 portion and the resultant suspension was refluxed for 30 min. Benzyl chloride (6.1 g, 48 mmol) was then added and the reaction mixture was refluxed for a further 48 h. On cooling, water (10 mL) was added and the solvent was removed in vacuo. The residue was treated with 6 N KOH until pH 12 was reached and then extracted with CH$_2$Cl$_2$. The organic phase was washed with brine, dried over MgSO$_4$, filtered, and concentrated in vacuo to give a yellow oil which on column chromatography over silica gel yielded the benzyl ether (8.55 g, 91%).

Reference: Patel, S. K.; Murat, K.; Py, S.; Vallée, Y. *Org. Lett.* **2003**, *5*, 4081–4084.

*Cleavage*

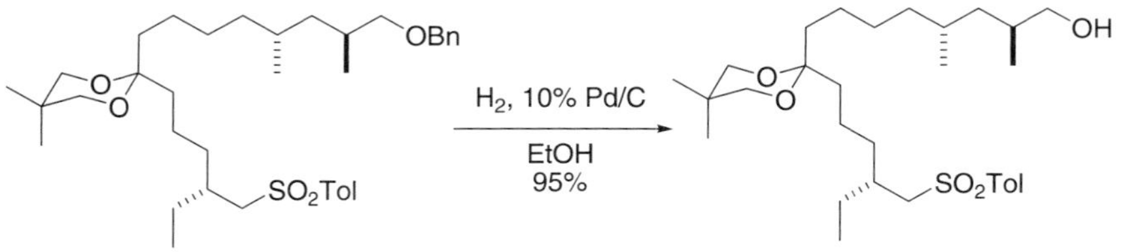

To a solution of the benzyl ether (762 mg, 1.44 mmol) in ethanol (10 mL) was added 10% palladium on activated carbon catalyst (50 mg). After stirring under an atmosphere of hydrogen for 2 h, the reaction mixture was filtered through a pad of Celite and concentrated to give the corresponding alcohol (603 mg, 95%) as a colorless oil.

Reference: Li, J., Ph.D. Thesis, *Total Synthesis of Myxovirescin A and Approaches Toward the Synthesis of the A/B Ring System of Zoanthamine*; Indiana University: Bloomington, Indiana, 1996; p. 174.

### 6.1.4 *para*-Methoxybenzyl

*para*-Methoxybenzyl (PMB) is a variant of the benzyl ether protective group. It can be differentiated from the benzyl ether because its cleavage can be accomplished by oxidation using ceric ammonium nitrate (CAN) or quinone.

*Protection*

To a solution of the cinnamyl alcohol (2.25 g, 8.75 mmol) in THF (100 mL) was added NaH (60% in mineral oil, 0.43 g, 10.5 mmol) followed by *p*-methoxybenzyl chloride, dimethylsulfoxide (5 mL), and a catalytic amount of *n*-Bu$_4$NI (0.20 g). The resulting suspension was stirred at room temperature overnight and quenched by slow addition of a saturated aqueous solution of NH$_4$Cl. After removal of THF in vacuo, the aqueous layer was extracted with CH$_2$Cl$_2$ (3 × 15 mL). The combined organic layers were washed with brine (10 mL), dried (Na$_2$SO$_4$), filtered, and concentrated in vacuo. Purification of the residue by flash chromatography eluting with EtOAc/hexane (1:9) afforded the primary alcohol (2.00 g, 89%) as a yellow oil.

*Cleavage*

To a solution of the PMB ether (30 mg, 0.073 mmol) in acetonitrile (2 mL) was added CAN (80 mg, 0.15 mmol). The resulting yellow solution was stirred at room temperature overnight and then quenched by addition of a saturated aqueous solution of NaHSO$_3$. After removal of acetonitrile, the aqueous layer was extracted with EtOAc (3 × 15 mL). The combined organic layers were washed with brine (10 mL), dried (MgSO$_4$), filtered, and concentrated in vacuo. Purification of the residue by flash chromatography eluting with EtOAc/hexane (2:1) afforded the diol (18 mg, 86%) as white crystals.

Reference: Li, J., Ph.D. Thesis, *Total Synthesis of Myxovirescin A and Approaches Toward the Synthesis of the A/B Ring System of Zoanthamine*. Indiana University: Bloomington, Indiana, 1996; p. 174.

## 6.1.5 Methyl ether

The methyl ether is one of the most popular protecting groups for phenols. The resulting phenoxy methyl ether, however, sometimes requires harsh conditions for cleavage.

*Protection*

To a stirred mixture of 3´,4´-dihydroxyacetophenone (100 mg, 0.66 mmol) and anhydrous K$_2$CO$_3$ (5 g, 36 mmol) in dry acetone (10 mL) was added MeI (1 mL, 16 mmol). The mixture was heated at reflux for 45 min, cooled to room temperature, filtered, and evaporated under reduced pressure. The residue was dissolved in CH$_2$Cl$_2$, washed with 2 portions of water, dried over anhydrous Na$_2$SO$_4$, and evaporated under reduced pressure. The product was purified by flash chromatography (silica, hexane/EtOAc 4:1, v/v). Removal of the solvent gave a 93% yield of the product, 3´,4´-dimethoxyacetophenone (110 mg, 0.61 mmol).

Reference: Khatib, S.; Nerya, O.; Musa, R.; Shmuel, M.; Tamir, S.; Vaya, J. *Bioorg. Med. Chem.* **2005**, *13*, 433–441.

*Cleavage*

A solution of the bis-dimethyl ether (0.5 mmol) in dry CH$_2$Cl$_2$ (3 mL) was cooled to –78 °C and then BBr$_3$ (0.2 mL, 2 mmol) was added. After the mixture was stirred for 30 min at –78 °C, the cooling bath was removed and the stirring was continued for an additional 24 h at room temperature. The reaction was quenched by addition of water (3 mL), and the CH$_2$Cl$_2$ was removed in vacuo. The aqueous solution was neutralized by addition of aqueous NaOH and then extracted with EtOAc (3 × 10 mL). The organic layer was washed with brine and dried over Na$_2$SO$_4$. After removal of the solvent, the residue was purified by column chromatography on silica with EtOAc/hexane (30–50%) as eluent to give the corresponding diol at 85% yield.

Reference: Liu, Y.; Ding, K. *J. Am. Chem. Soc.* **2005**, *127*, 10488–10489.

## 6.1.6 Methoxymethylene ether

The methoxymethylene (MOM) ether belongs to a class of substituted ether protecting groups. Unlike methyl ether, however, the MOM ether is actually an acetal, which is cleaved under acidic conditions. The MOM ether is the most robust among all alkoxymethyl ether protecting groups.

*Protection*

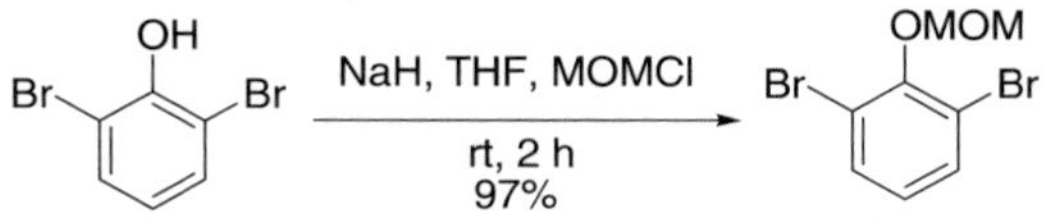

[Scheme: 2,6-dibromophenol (OH between two Br on benzene) → with NaH, THF, MOMCl, rt, 2 h, 97% → 2,6-dibromo-OMOM arene]

To a solution of 2,6-dibromophenol (58.0 g, 230.2 mmol) in THF (300 mL) at 0 °C was added NaH (13.8 g, 60% dispersion in mineral oil). After 5 min of stirring at 0 °C, MOMCl (22.5 mL, 300 mmol) was added via a syringe. The reaction mixture was then allowed to slowly warm to 25 °C, and following 2 h of additional stirring at that temperature the reaction mixture was poured into $Et_2O$ (300 mL) and washed extensively with 3 M NaOH (3 × 75 mL) to remove any residual phenol starting material. The organic layer was then dried ($MgSO_4$) and concentrated to give MOM-protected 2,6-dibromophenol (66.2 g, 97%) as a yellow oil.

Reference: Nicolaou, K. C.; Snyder, S. A.; Huang, X.; Simonsen, K. B.; Koumbis, A. E.; Bigot, A. *J. Am. Chem. Soc.* **2004**, *126*, 10162–10173.

*Cleavage*

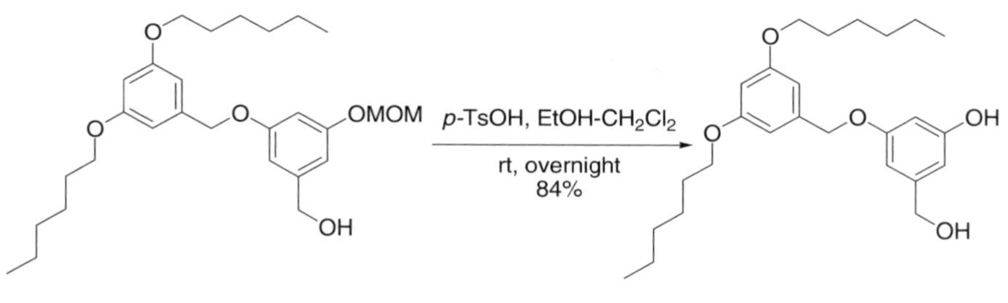

[Scheme: dendrimer with OMOM and OH → p-TsOH, EtOH-$CH_2Cl_2$, rt, overnight, 84% → dendrimer with two OH groups]

The MOM ether (0.63 g, 1.32 mmol) was dissolved in a mixture of ethanol and minimal amount of $CH_2Cl_2$. To this solution, *p*-toluenesulfonic acid (*p*-TsOH, 3–6 equivalents) was added and the reaction mixture was stirred at room temperature overnight. The reaction was then refluxed for 1 h to promote complete conversion, after which the solvent was removed in vacuo and the residue was partitioned between water and $CH_2Cl_2$. The organic layer was dried over $Na_2SO_4$ and evaporated to yield the crude product, which was then purified by silica gel chromatography using EtOAc/hexane (2:8) as the eluent to give the corresponding phenol (0.48 g, 84%).

Reference: Sivanandan, K.; Aathimanikandan, S. V.; Arges, C. G.; Bardeen, C. J.; Thayumanavan, S. *J. Am. Chem. Soc.* **2005**, *127*, 2020–2021.

### 6.1.7 *tert*-Butyldiphenylsilyl ether

*tert*-Butyldiphenylsilyl (TBDPS) protection is one of the most popular alcohol protective groups. It is more stable than other silyl ethers under acidic conditions.

PROTECTING GROUPS 175

TBDPS ether is visible under UV light and thus it is more advantageous than the corresponding TBDMS ether if there is no chromophore present in the alcohol. Selective protection of the primary alcohol can be achieved as shown below. Silyl chloride and cleavage by-product silylol are both lachrymators and therefore the reactions should be carried out in the hood. The procedures for protection and cleavage are applicable to other silyl ethers such as *tert*-butyldimethylsilyl, trimethylsilyl, and tri-isopropylsilyl.

*Protection*

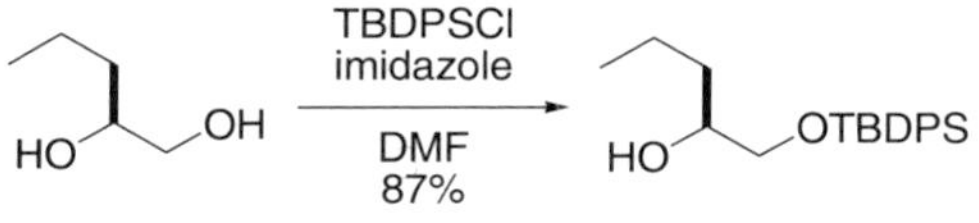

HO~~~OH  →(TBDPSCl, imidazole, DMF, 87%)→  HO~~~OTBDPS

To a solution of (2S)-pentane-1,2-diol (86 mg, 0.83 mmol) in dimethylformamide (DMF, 95 mL) was added imidazole (84 mg, 1.2 mmol) followed by *tert*-butyldiphenylsilyl chloride (227 mg, 0.83 mmol, 0.22 mL). The resulting solution was stirred for 1 h, diluted with diethyl ether (40 mL), washed with water (3 × 5 mL) and brine (10 mL), dried (MgSO$_4$), filtered, and concentrated in vacuo. Purification of the residue by flash chromatography eluting with EtOAc/hexane (1:9) afforded (2S)-1-(*tert*-butyldiphenylsilanyloxy)-pentan-2-ol (247 mg, 87%) as a colorless oil.

Reference: Li, J., Ph.D. Thesis, *Total Synthesis of Myxovirescin A and Approaches Toward the Synthesis of the A/B Ring System of Zoanthamine*. Indiana University: Bloomington, Indiana, 1996; p. 174.

*Cleavage*

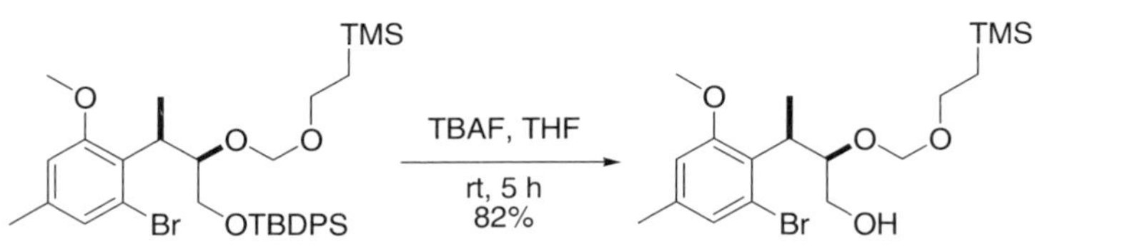

[Structure with OTBDPS] →(TBAF, THF, rt, 5 h, 82%)→ [Structure with OH]

To a solution of the TBDPS ether (180 mg, 0.38 mmol) in THF (4 mL) was added a tetrabutylammonium fluoride solution (1.0 M in THF, 0.76 mL, 0.76 mmol). The resulting yellow solution was stirred at room temperature for 5 h and quenched by addition of a saturated aqueous solution of NH$_4$Cl. After removal of THF in vacuo, the aqueous layer was extracted with EtOAc (3 × 15 mL). The combined organic layers were washed with brine (10 mL), dried (MgSO$_4$), filtered, and concentrated in vacuo. Purification of the residue by flash chromatography eluting with EtOAc/hexane (1:9) afforded the primary alcohol (73 mg, 82%) as a colorless oil.

Reference: Li, J., Ph.D. Thesis, *Total Synthesis of Myxovirescin A and Approaches Toward the Synthesis of the A/B Ring System of Zoanthamine*. Indiana University: Bloomington, Indiana, 1996; p. 174.

### 6.1.8 Tetrahydropyran

Tetrahydropyranyloxy (THP) is an often-used and inexpensive protective group. The greatest disadvantage of this protecting group is that it creates an additional chiral center which can make the NMR spectra more complicated.

*Protection*

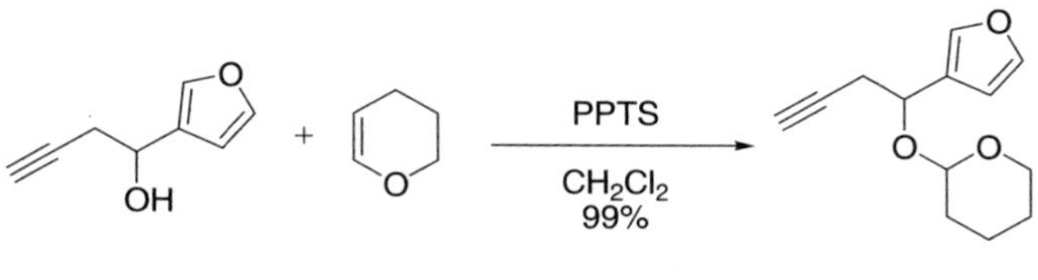

A mixture of 1-(3-furyl)but-3-yn-1-ol (500 mg, 3.68 mmol), $CH_2Cl_2$ (10 mL), dihydropyran (0.67 mL, 7.4 mmol), and PPTS (90 mg, 0.37 mmol) was stirred until the disappearance of the starting material. The solution was concentrated under vacuum and the crude product was purified by flash chromatography (light petroleum–$Et_2O$, 7/3) to give 1-(3-furyl)-1-(2-tetrahydropyranyl-oxy)but-3-yne, as a 1/1 mixture of diastereoisomers, at 99% yield.

Reference: Commeiras, L.; Parrain, J.-L. *Tetrahedron: Asymmetry* **2004**, *15*, 509–517.

*Cleavage*

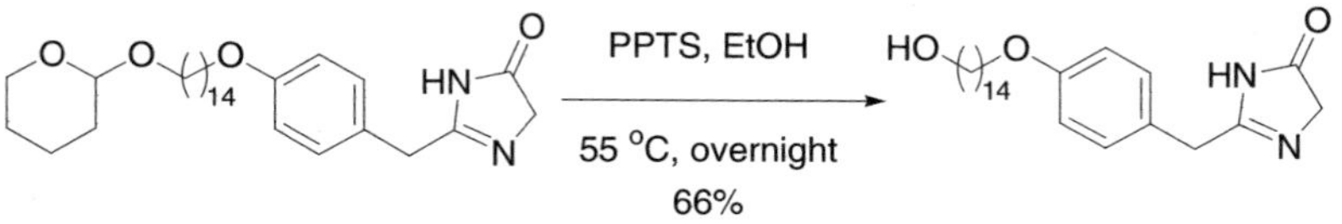

The THP starting material (1.80 g) and PPTS (76 mg, 0.30 mmol) were heated at 55 °C in EtOH (30 mL) overnight. Water (about 40 mL) was added to the mixture while it was maintained at reflux overnight. The clear yellow solution was then kept at room temperature and the insoluble brown oil was separated from the supernatant by decantation. After drying under reduced pressure, the brown oil was purified by silica gel column chromatography (MeOH/$CH_2Cl_2$) and then recrystallized ($CHCl_3$) to afford the alcohol as white crystals (0.80 g, 66% yield).

Reference: Dong, C.-Z.; Ahamada-Himidi, A.; Plocki, S.; Aoun, D.; Touaibia, M.; Meddad-Bel Habich, N.; Huet, J.; Redeuilh, C.; Ombetta, J.-E.; Godfroid, J.-J.; Massicot, F.; Heymans, F. *Bioorg. Med. Chem.* **2005**, *13*, 1989–2007.

## 6.2 Amines and Anilines

In this section, the protections and cleavages of seven protecting groups for amines and anilines are presented: benzyl; BOC; Cbz; Fmoc; phthaloyl; sulfonamide; and trifluoroacetyl.

## 6.2.1 Benzyl

While the benzyl ether is readily removed via hydrogenolysis, the corresponding benzyl protected of amines is less readily cleaved.

*Protection*

$$\text{Cl}^{\ominus} \ \text{H}_3\overset{\oplus}{\text{N}}\text{-CH(CH}_2\text{Ph)-CO}_2\text{Me} \quad \xrightarrow[\text{95 °C, 19 h}]{\text{BnCl, Na}_2\text{CO}_3\text{, water}} \quad \text{Bn}_2\text{N-CH(CH}_2\text{Ph)-CO}_2\text{Me}$$

85%

To a solution of *L*-phenylalanine methyl ester hydrochloride (25.0 g, 151 mmol) and Na$_2$CO$_3$ (66.7 g, 483 mmol) dissolved in water (100 mL) was added benzyl chloride (57.5 g, 454 mmol), and the mixture was heated at 95 °C with stirring for 19 h. After the reaction mixture was cooled to ambient temperature, water (50 mL) and *n*-heptane (67 mL) were added and extracted. The organic layer was separated and washed twice with a mixed solution of methanol/water (1:2, 50 mL) and then dried over anhydrous Na$_2$SO$_4$. Concentration of the solution in vacuo provided Bn$_2$-*L*-Phe-OMe (61.6 g, 85%) as a colorless oil.

Reference: Suzuki, T.; Honda, Y.; Izawa, K.; Williams, R. M. *J. Org. Chem.* **2005**, *70*, 7317–7323.

*Cleavage*

$$\text{Bn}_2\text{N-CH(CH}_2\text{Ph)-CH(OH)-CO}_2\text{H} \quad \xrightarrow[\text{MeOH, rt}]{\text{H}_2\text{, Pd/C, AcOH}} \quad \text{H}_2\text{N-CH(CH}_2\text{Ph)-CH(OH)-CO}_2\text{H}$$

97%

To a solution of the dicyclohexylamine salt of the dibenzylamine (2.8 g, 5.0 mmol) dissolved in a mixture of methanol (25 mL) and acetic acid (2.4 mL) was added 5% palladium on carbon (water content 53.3%, 1.2 g, 0.23 mmol Pd). The resulting mixture was stirred for 25 h under hydrogen at atmospheric pressure and ambient temperature. NaOH (2 M) aqueous solution (about 20 mL) was added to the reaction mixture in a water bath to adjust the pH to 5.1 at 30 °C. After the mixture was stirred for 40 min at ambient temperature, it was filtered to remove the catalyst. Filtration and concentration afforded the cleaved amine (949 mg, 97%). Recrystallization using EtOAc gave the sodium salt as white crystals (602 mg, 57% as a recovered yield in crystallization).

Reference: Suzuki, T.; Honda, Y.; Izawa, K.; Williams, R. M. *J. Org. Chem.* **2005**, *70*, 7317–7323.

## 6.2.2 t-Butyloxycarbonyl

t-Butyloxycarbonyl (BOC) is the most widely used protecting group for amines.

*Protection*

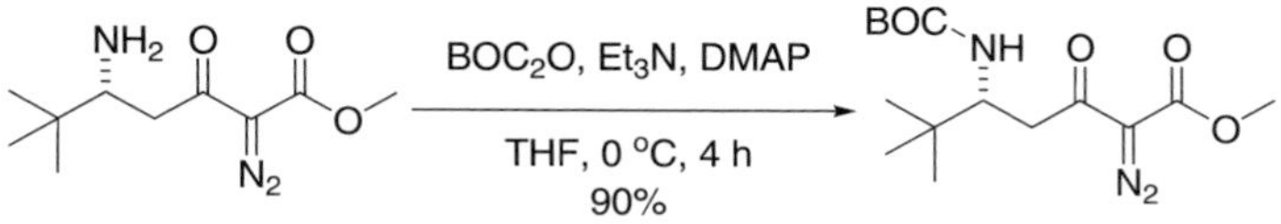

To a solution of the amine (1.4 g, 4.1 mmol) in THF (20 mL) at 0 °C was added Et₃N (3.4 mL, 2.5 g, 24.6 mmol), followed by DMAP (about 0.01 g). To the reaction mixture was added di-*tert*-butyl dicarbonate (0.95 g, 4.92 mmol) and the solution was stirred at this temperature for 4 h. At this time the reaction mixture was quenched with ice and water (30 mL) and extracted with EtOAc (2 × 30 mL). The combined organic phases were washed with H₂O (20 mL) and brine (20 mL), dried (Na₂SO₄), and concentrated. Chromatography (5%–30% EtOAc/hexane) afforded the BOC-amine as a white solid (1.2 g, 90%).

Reference: Davis, F. A.; Yang, B.; Deng, J. *J. Org. Chem.* **2003**, *68*, 5147–5152.

*Cleavage*

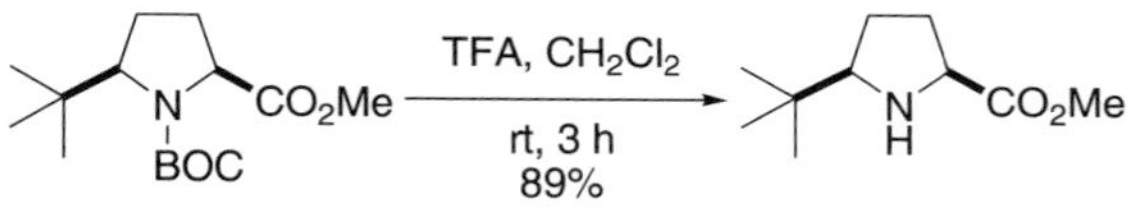

In a 5-mL round-bottomed flask equipped with a stir bar and a rubber septum under an argon atmosphere was placed the BOC-pyrrolidine (0.033 g, 0.12 mmol) in CH₂Cl₂ (2 mL). The solution was cooled to 0 °C, and TFA (180 μL, 2.31 mmol) was added. The reaction mixture was stirred at room temperature for 3 h, quenched with saturated aqueous NaHCO₃ (2 mL), and stirred for 20 min. The organic phase was extracted with CH₂Cl₂ (3 × 3 mL). The combined organic phases were washed with brine (5 mL), dried (Na₂SO₄), and concentrated. Chromatography (1:1 EtOAc/hexane) afforded the pyrrolidine as a yellow low-melting solid (0.019 g, 89%).

Reference: Davis, F. A.; Yang, B.; Deng, J. *J. Org. Chem.* **2003**, *68*, 5147–5152.

## 6.2.3 2,5-Dimethylpyrrole

The 2,5-dimethylpyrrole protecting group is a robust blocking group for the diprotection of primary amines. The protecting group is stable to a variety of nucleophiles (i.e., RLi, RMgX, etc.), reducing agents (i.e., LiAlH₄), and low reactivity toward electrophiles (i.e., acid chlorides). Introduction of the protecting group is straightforward where an amine and acetonylacetone are condensed to form the pyrrole. The condensation is accelerated in the presence of acetic acid.

*Protection*

A mixture of 5-aminoindole (120.0 g, 0.908 mol), acetonylacetone (200.0 mL, 1.70 mol), and toluene (400 mL) was heated at reflux under nitrogen using a Dean-Stark trap for 6 h. The reaction was cooled and then poured through a silica gel filter (~ 2 kg) followed first by hexanes (4 L) and then by 6% ether in hexanes to afford 133.3 g of a pink solid. Recrystallization of this solid in ether/hexanes afforded 126.1 g (66%) of the 2,5-dimethylpyrrole as an off-white solid.

Reference: Macor, J. E.; Chenard, B. L.; Post, R. J. *J. Org. Chem.* **1994**, *59*, 7496–7498.

*Cleavage*

Cleavage of the 2,5-dimethylpyrrole has been achieved using $NH_2OH \cdot HCl$/base, $TFA/H_2O$, or singlet oxygen. The deprotection is normally done using $NH_2OH \cdot HCl$/$Et_3N$ or KOH in a refluxing mixture of *i*-PrOH or EtOH and water. Despite the relatively harsh conditions for the removal of this protecting group, the cleavage has been achieved in highly functionalized molecules. See (a) Bowers, S. G.; Coe, D. M.; Boons, G.-J. *J. Org. Chem.* **1998**, *63*, 4570–4571. (b) Baker, R.; Castro, J. L. *J. Chem. Soc. Perkin Trans. 1* **1990**, 47–65.

A mixture of the 2,-5-dimethylpyrrole (81.5 g, 0.265 mol), hydroxylamine hydrochloride (368 g, 5.30 mol), and triethylamine (367 mL, 2.65 mol) in 2-propanol (800 mL) and water (200 mL) was heated at reflux under nitrogen for 4.5 h. The resulting reaction mixture was cooled in an ice bath, solid sodium hydroxide (212 g, 5.30 mol) was added, and the resulting reaction mixture was stirred at room temperature under nitrogen for 24 h. The reaction mixture was then filtered through Celite, and the filtrate was evaporated under reduced pressure. The residual oil was passed

through a silica gel filter (~ 1 kg) followed by elution with EtOAc/MeOH/Et$_3$N (8:1:1) to afford 85 g of a pale yellow solid. The solid was dissolved in EtOAc (1 L), and this solution was washed with a saturated solution of sodium chloride (3 × 100 mL). The organic layer was dried (Na$_2$SO$_4$) and evaporated under reduced pressure to afford 50.55 g (83%) of the 5-aminoindole.
Reference: Macor, J. E.; Chenard, B. L.; Post, R. J. *J. Org. Chem.* **1994**, *59*, 7496–7498.

### 6.2.4 Carbobenzyloxy

Carbobenzyloxy (Cbz, *O*-benzyloxycarbonyl) may be removed by either hydrogenation or acid (e.g., TFA).

*Protection*

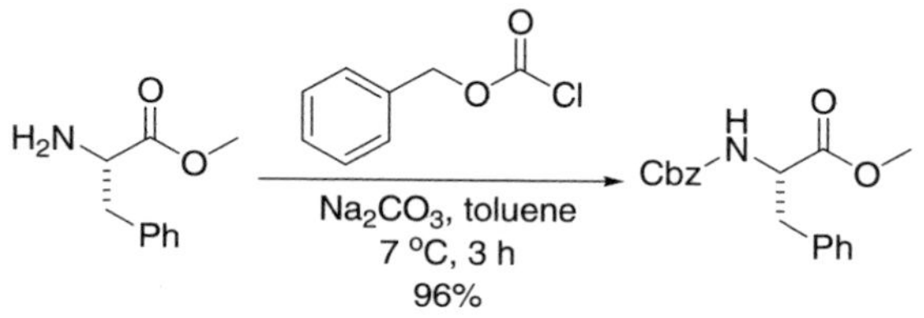

To a mixture of *L*-phenylalanine methyl ester hydrochloride (20.0 g, 93 mmol) suspended in toluene (93 mL) was added benzyl chloroformate (15.8 g, 93 mmol) with cooling in an ice bath. An aqueous solution of Na$_2$CO$_3$ (1 M, 130 mL) was added dropwise with vigorous stirring at 7 °C or lower. After this addition was complete, the mixture was stirred for 3 h. The organic layer was separated, washed with 0.1 M HCl (60 mL) and saturated NaHCO$_3$ solution (60 mL), and then dried over anhydrous Na$_2$SO$_4$. Concentration of the solution in vacuo provided Cbz-*L*-Phe-OMe (28.8 g, 96%) as a colorless oil.
Reference: Suzuki, T.; Honda, Y.; Izawa, K.; Williams, R. M. *J. Org. Chem.* **2005**, *70*, 7317–7323.

*Cleavage*

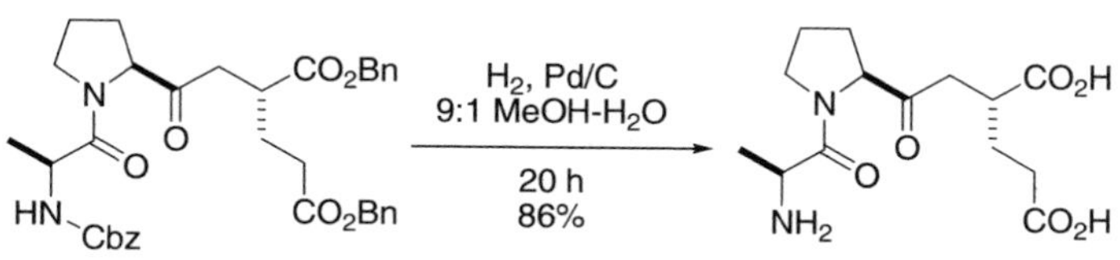

A mixture of the above-protected peptide (0.30 g, 0.48 mmol) and 10 wt% palladium on activated carbon (0.05 g, 0.13 mmol) in methanol (40 mL) was stirred under an atmosphere of hydrogen at room temperature for 20 h. The solution was filtered through a Celite pad and the pad was washed with methanol (2 × 25 mL). The filtrate was evaporated to dryness, dissolved in methanol (35 mL), and re-filtered through a Celite pad. The solution was evaporated to dryness, dried in vacuo, and triturated with anhydrous ether to give D-alanyl-L-prolyl-L-glutamic acid (0.13 g, 86%) as a white solid.

Reference: Lai, M. Y. H.; Brimble, M. A.; Callis, D. J.; Harris, P. W. R.; Levi, M. S.; Sieg, F. *Bioorg. Med. Chem.* **2005**, *13*, 533–548.

### 6.2.5 9-Fluorenylmethyl carbamate

9-Fluorenylmethyl carbamate (Fmoc) is widely used in peptide chemistry and solid-phase peptide chemistry. Fmoc is often uniquely cleaved by amines and is stable under acidic conditions.

*Protection*

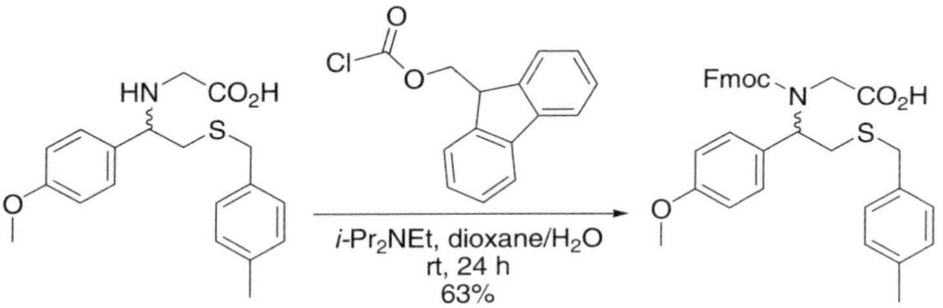

To a stirring solution of the amino acid (86.3 mg, 0.25 mmol) in a mixture of dioxane/$H_2O$ (2:1, 3 mL) was added diisopropylethylamine (2.5 mmol) followed by a dioxane solution of Fmoc-Cl (1.1 mmol in 0.65 mL dioxane). The reaction mixture was stirred at room temperature for 24 h, poured into water (5 mL), and extracted with ether. The aqueous layer was then acidified with 1 M HCl solution and extracted with EtOAc. The combined organic layer was dried ($MgSO_4$), filtered, and concentrated. Flash chromatography on silica gel (EtOAc/pentane, 30% to 50%) afforded the Fmoc protected amino acid as a white powder (89 mg, 63%).

Reference: Tchertchian, S.; Hartley, O.; Botti, P. *J. Org. Chem.* **2004**, *69*, 9208–9214.

*Cleavage*

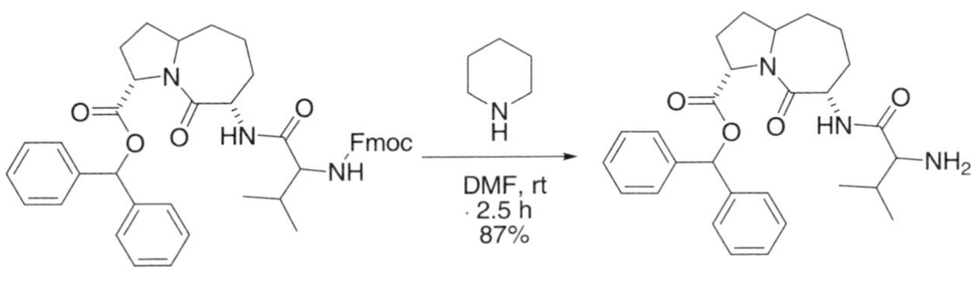

The Fmoc-protected peptide (310 mg, 0.44 mmol) was dissolved in dry DMF (6 mL). Piperidine (1.2 mL, 20% by volume) was added and the solution was stirred at room temperature under nitrogen for 2.5 h. The solvent was removed in vacuo at < 50 °C to yield a white solid, which was purified by column chromatography on silica gel, eluting with a gradient of methanol–dichloromethane (1:99 to 5:95). The deprotected peptide was collected as a yellow foam (182 mg, 87%).

Reference: Davies, D. E.; Doyle, P. M.; Hill, R. D.; Young, D. W. *Tetrahedron* **2005**, *61*, 301–312.

### 6.2.6 Phthaloyl

Phthalimides are relatively stable under both acidic and basic conditions but are easily cleaved by nucleophiles. Its cleavage is often accomplished using hydrazine according to the Ing–Manske procedure of the Gabriel reaction.

*Protection*

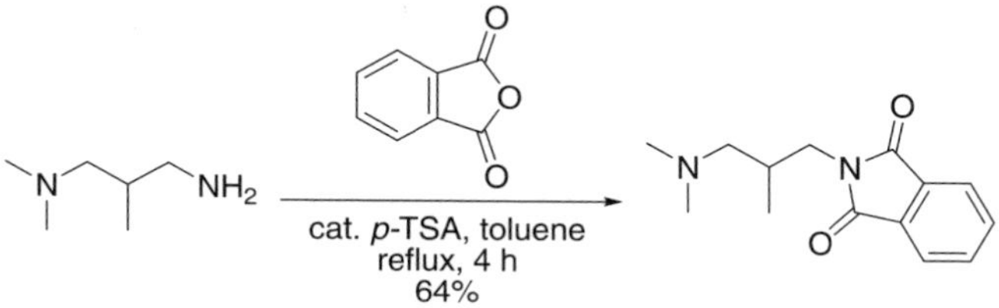

A mixture of phthalic acid anhydride (40 mmol), $N^1,N^1$,2-trimethylpropane-1,3-diamine (4.6 g, 40 mmol), and a catalytic amount of *p*-toluenesulfonic acid in toluene (100 mL) was refluxed using a water separator. After a reaction time of 4 h, the solvent was evaporated. The oily residues were purified by means of column chromatography (silica gel, eluent $CH_2Cl_2$:MeOH = 1:1). The obtained oils crystallized after a few hours at room temperature and the product was obtained at 64% yield.

Reference: Muth, M.; Sennwitz, M.; Mohr, K.; Holzgrabe, U. *J. Med. Chem.* **2005**, *13*, 2212–2217.

*Cleavage*

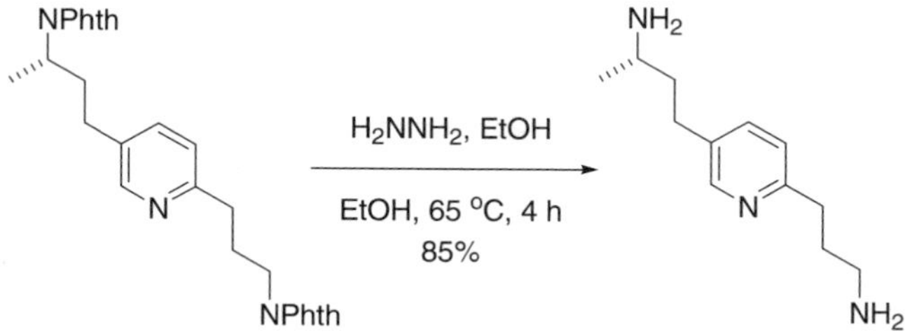

A mixture of the bis-phthaloyl starting material (16.6 g, 35.5 mmol), hydrazine hydrate (10.2 mL, 328 mmol), and ethanol (400 mL) was heated at 65 °C for 4 h. The mixture was filtered, and the cake was washed with toluene and azeotropically dried. The diamine was obtained at 85% yield.

Reference: Hartner, F. W.; Hsiao, Y.; Palucki, M. *J. Org. Chem.* **2004**, *69*, 8723–8730.

## 6.2.7 Sulfonamide

While sulfonamides are robust protecting groups for amines and anilines, their cleavage often calls for dissolving metal chemistry.

*Protection*

The diamine (2.99 g, 11.1 mmol) and Et$_3$N (1.7 mL, 12.2 mmol) were dissolved in CH$_2$Cl$_2$ (50 mL) and cooled in an ice bath. *p*-Toluenesulfonyl chloride (2.29 g, 12.0 mmol) was added slowly to the solution. The reaction mixture was stirred for 2 h, and then saturated aqueous NaHCO$_3$ (50 mL) was added and the solution was stirred overnight to quench excess sulfonic acid. The aqueous solution was drawn off and the CH$_2$Cl$_2$ layer was washed (2 × 50 mL) with saturated aqueous NaHCO$_3$ and water. The organic solution was dried (Mg$_2$SO$_4$) and decanted, and the solvent was removed in vacuo. The resulting amorphous white solid was recrystallized from EtOAc: hexane (3:1) to give the mono-protected product as colorless crystals (4.49 g, 96%).

Reference: Goodwin, J. M.; Olmstead, M. M.; Patten, T. E. *J. Am. Chem. Soc.* **2004**, *126*, 14352–14353.

*Cleavage*

A flask was charged with the bis-sulfonamide (1.31 g, 3.00 mmol), and NH$_3$ was condensed at –78 °C with magnetic stirring. The mixture was allowed to warm to –33 °C and solid Li (220 mg, 31.7 mmol) was gradually added. The suspension turned blue and the reaction was quenched after 80 min by dropwise addition of brine (0.5 mL). NH$_3$ was evaporated and the residue was diluted with H$_2$O (10 mL) and concentrated in vacuo until the condensation of water was observed. The residue was diluted with water (10 mL), acidified with concentrated HCl, and extracted with CH$_2$Cl$_2$. The aqueous phase was concentrated in vacuo until a white precipitate formed. NaOH (25% aqueous, 15 mL) was added and the aqueous solution was extracted with CH$_2$Cl$_2$ in a liquid/liquid extractor for 42 h. Concentration of the organic phase yielded 2,5-diaminobicyclo[2.2.1]heptane as a colorless oil (321 mg, 85%).

Reference: Berkessel, A.; Schroeder, M.; Sklorz, C. A.; Tabanella, S.; Vogl, N.; Lex, J.; Neudoerfl, J. M. *J. Org. Chem.* **2004**, *69*, 3050–3056.

### 6.2.8 Trifluoroacetyl

Trifluoroacetyl amides are an easily installed and cleaved protecting group for amines and anilines.

*Protection*

A dry, 3-L, three-necked, round-bottomed flask equipped with a mechanical stirrer and a 200-mL addition funnel was charged with the HCl salt of the starting material amino ester (21.46 g, 92.7 mmol) and $CH_2Cl_2$ (134 mL). The solution was stirred while $Et_3N$ (19.69 g, 195 mmol) was added in 1 portion. The solution was cooled to −50 °C and trifluoroacetic anhydride (21.41 g, 102 mmol) was added dropwise to the reaction mixture through the addition funnel over 1 h. The solution was stirred at −50 °C for 1 h and then allowed to warm to 0 °C. Aqueous HCl (1.5%, 100 mL) was added to the solution in 1 portion, and the aqueous layer was separated by extraction. The organic extracts were washed with $H_2O$ (100 mL), dried ($Na_2SO_4$), filtered, and concentrated in vacuo to give a pale yellow solid. The solid was dissolved in EtOAc and the resulting solution was purified by column chromatography ($SiO_2$, EtOAc/hexane, 1:1) to yield the trifluoroacetyl amide as a white solid (21 g). The solid was recrystallized from $CH_2Cl_2$/hexane to yield a white crystalline solid (17.6 g, 65%).

Reference: Itagaki, M.; Masumoto, K.; Yamamoto, Y. *J. Org. Chem.* **2005**, *70*, 3292–3295.

*Cleavage*

A suspension of trifluoroacetyl amide (158 mg, 0.434 mmol) and potassium carbonate (310 mg, 2.25 mmol) in methanol–water (12:1, 13 mL) was heated at reflux for 2 h. The solvent was removed in vacuo, the residue was dissolved in dichloromethane (30 mL), and water (20 mL) was added. The aqueous layer was extracted with a further 2 portions of dichloromethane (30 mL). The combined organic fractions were dried ($MgSO_4$) and the solvent was removed in vacuo to yield the deprotected amine as a yellow oil (100 mg, 86%).

Reference: Davies, D. E.; Doyle, P. M.; Hill, R. D.; Young, D. W. *Tetrahedron* **2005**, *61*, 301–312.

## 6.3 Aldehydes and Ketones

In this section, the protections and cleavages of four protecting groups for aldehydes and ketones are presented: dimethylketal; 1,3-dioxane; 1,3-dioxolane; and 1,3-dithiane.

### 6.3.1 Dimethylketal

Protection of a ketone as the corresponding dimethylketal is an older method. Dimethylketal are more readily cleaved under weakly acidic conditions compared with cyclic ketals.

*Protection*

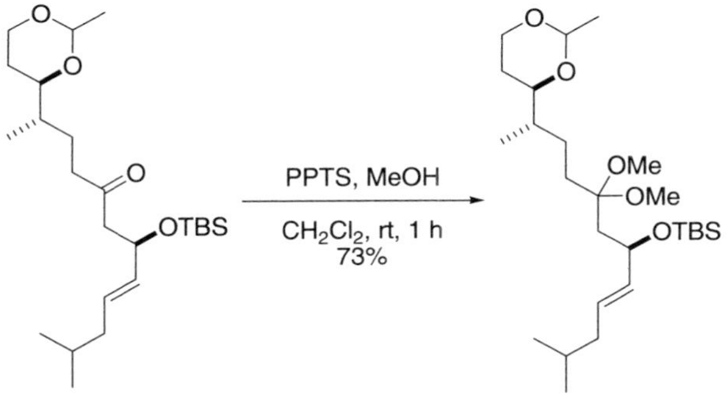


The ketone (0.33 g, 0.64 mmol) was diluted with anhydrous MeOH (7 mL) and catalytic amount of PPTS (0.008 g, 0.032 mmol) was added. The reaction mixture was stirred for 1 h and then diluted with $CH_2Cl_2$ and work up with saturated $NaHCO_3$ solution. The layers were separated and the aqueous layer was extracted with $CH_2Cl_2$ (2 × 15 mL). The combined organic fractions were dried over $Na_2SO_4$, filtered, and concentrated in vacuo. Purification by flash chromatography provided the dimethyl ketal (0.22 g, 73%).

Reference: Crimmins, M. T.; Siliphaivanh, P. *Org. Lett.* **2003**, *5*, 4641–4644.

*Cleavage*

Even brief exposure to aqueous HCl is enough to cleave the dimethyl acetal, thus revealing the ketone functionality.

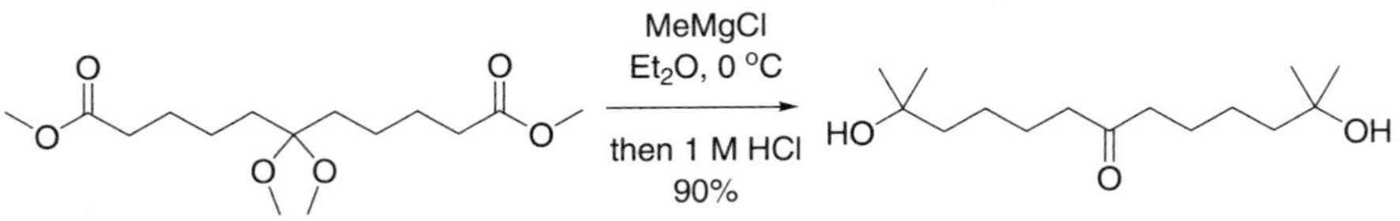

At 0 °C under a $N_2$ atmosphere, MeMgCl (22% (w/w) in THF (41.5 mL, 0.126 mol) was added dropwise to a solution of the bis-ester (8.5 g, 90% pure, 25.2 mmol) in $Et_2O$ (100 mL) for 30 min. After stirring for 30 min, the reaction mixture was allowed to warm to room temperature, stirred for 2.5 h, and then cooled again to 0 °C. The reaction was quenched by careful addition of HCl (1 M, 125 mL) and the layers were separated. The aqueous phase was extracted with $Et_2O$ (50 mL) and the combined organic layers were washed with brine (2 × 25 mL) and dried. The remaining residue was purified by column chromatography (silica, EtOAc) to give 6.91 g of a brown oil, which was taken up in EtOAc (25 mL). Norrit (0.5 g) was added and the suspension was filtered through kieselguhr and washed with EtOAc (50 mL). The combined filtrate and washings were evaporated in vacuo to give 2,12-dihydroxy-2,12-dimethyl-7-tridecanone (6.73 g, 93%) as a dark yellow oil.

Reference: Bell, R. P. L.; Verdijk, D.; Relou, M.; Smith, D.; Regeling, H.; Ebbers, E. J.; Leemhuis, F. M. C.; Oniciu, D. C.; Cramer, C. T.; Goetz, B.; Pape, M. E.; Krauseand, B. R.; Dasseux, J.-L. *Bioorg. Med. Chem.* **2004**, *12*, 223–236.

### 6.3.2 1,3-Dioxane

1,3-Dioxane is one of the most popular protecting groups for carbonyls.

*Protection*

To a solution of the ynone (3.8 g, 28.3 mmol) in benzene (70 mL) was added PPTS (0.7 g, 2.83 mmol) and 2,2-dimethyl-1,3-propane diol (2.95 g, 28.3 mmol), and the resulting solution was refluxed overnight under Dean–Stark conditions. The reaction was cooled to room temperature and quenched with saturated $NaHCO_3$ solution. The product was extracted with $CH_2Cl_2$ and the organic layer dried over $MgSO_4$ and the solvent evaporated. Purification of the residue by flash chromatography (hexane/$CH_2Cl_2$ 10:1 to 4:1) furnished the acetal (2.5 g, 48%).

Reference: Rigby, J. H.; Laxmisha, M. S.; Hudson, A. R.; Heap, C. H.; Heeg, M. J. *J. Org. Chem.* **2004**, *69*, 6751–6760.

*Cleavage*

Aqueous sulfuric acid (50%, 0.4 mL) was added to a solution of the starting material (635 mg, 0.91 mmol) in methanol (9 mL). The reaction mixture was stirred at room temperature for 2 h. It was then cautiously poured into saturated aqueous sodium bicarbonate solution. The aqueous layer was extracted with EtOAc. The organic portions were combined, washed with brine, dried over anhydrous $Na_2SO_4$, filtered, and evaporated. Chromatography of the residue over silica gel using hexane/EtOAc afforded the bis-ketone (428 mg, 92%) as a colorless foam.

Reference: Fuhrmann, U.; Hess-Stumpp, H.; Cleve, A.; Neef, G.; Schwede, W.; Hoffmann, J.; Fritzemeier, K.-H.; Chwalisz, K. *J. Med. Chem.* **2000**, *43*, 5010–5016.

### 6.3.3 1,3-Dioxolane

Among all cyclic ketals, five-membered cyclic ketals (1,3-dioxolanes) are more readily hydrolyzed then the corresponding six-membered cyclic ketals (1,3-dioxanes) under acidic conditions.

*Protection*

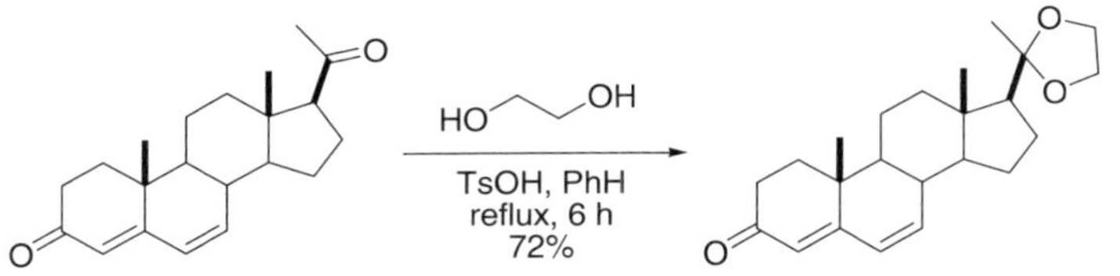

A mixture of pregna-4,6-diene-3,20-dione (2.70 g, 8.64 mmol), ethylene glycol (5.0 mL), *p*-TsOH (100 mg, 0.52 mmol) and benzene (150 mL) was stirred and heated under reflux for 6 h. Water formed during the reaction was removed by a Dean–Stark trap. The cooled reaction mixture was diluted with $Et_2O$ (600 mL) and sequentially washed with saturated aqueous $NaHCO_3$, water, and brine. The $Et_2O$ extract was stirred and dried over anhydrous $MgSO_4$ (35 g, 0.29 mol) for several hours until thin layer chromatography showed that the ketal group at C-3 was hydrolyzed completely. The mixture was filtered and the solvent was evaporated. The residue was purified by silica gel column chromatography to give the 1,3-dioxolane product (2.32 g, 75%) as a white solid.

Reference: Zeng, C.-M.; Manion, B. D.; Benz, A.; Evers, A. S.; Zorumski, C. F.; Mennerick, S.; Covey, D. F. *J. Med. Chem.* **2005**; *48*; 3051–3059.

*Cleavage*

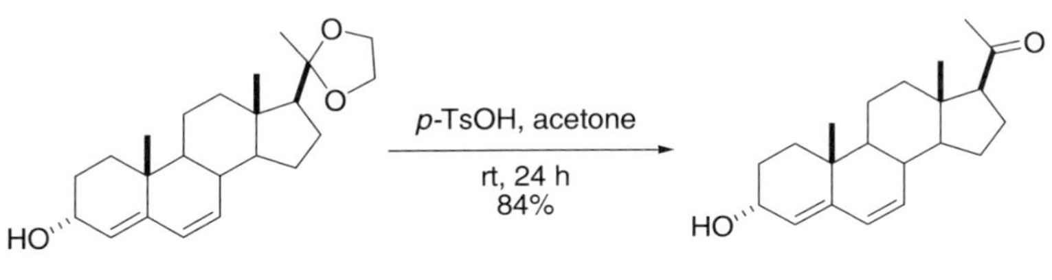

The ketal (70 mg, 0.19 mmol) was treated with *p*-TsOH (10.0 mg, 0.05 mmol) in acetone (15 mL). The mixture was stirred at room temperature for 24 h and extracted with Et$_2$O. The extract was washed with saturated aqueous NaHCO$_3$, water, and brine, and dried. The residue was purified by silica gel column chromatography to give (3α,5α,7α)-3-hydroxy-7-methylpregnan-20-one (52 mg, 84%) as a white solid.

Reference: Zeng, C.-M.; Manion, B. D.; Benz, A.; Evers, A. S.; Zorumski, C. F.; Mennerick, S.; Covey, D. F. *J. Med. Chem.* **2005**, *48*, 3051–3059.

### 6.3.4 1,3-Dithiane

1,3-Dithianes are very robust towards acidic hydrolysis. They have to be cleaved using mercury salts or oxidative procedure. One disadvantage is the stench of thiols, which may be alleviated by rinsing the reaction apparatuses with bleach.

*Protection*

A mixture of 4-chloro-1-(4-fluorophenyl)butan-1-one (15.0 g, 74.8 mmol), propane-1,3-dithiol (10.0 g, 92.4 mmol), and a 2.0 M ethereal hydrogen chloride solution (90 mL, 180 mmol of HCl) was stirred at 20 °C for 3 days and then cooled to 0 °C, followed by the addition of water (100 mL). The organic phase was separated, washed with water (2 × 100 mL), and dried over anhydrous Na$_2$SO$_4$. The solvent and the excess propane-1,3-dithiol were removed under reduced pressure (crystallization of the residue on standing at 20 °C), and the resulting solid was washed with methanol (50 ml) and then recrystallized from *n*-pentane (slow cooling of a saturated boiling solution to 20 °C) to give 17.8 g (82%) of 2-(3-chloropropyl)-2-(4-fluorophenyl)-1,3-dithiane as a colorless crystalline solid.

Reference: Heinrich, T.; Burschka, C.; Penka, M.; Wagner, B.; Tacke, R. *J. Organomet. Chem.* **2005**, *690*, 33–47.

*Cleavage*

To a mixture of the 1,3-dithiane starting material (1.70 g, 3.2 mmol), CaCO$_3$ (500 mg, 5.0 mmol), and THF (200 mL) at room temperature was added mercury (II) perchlorate (167 mL of 0.025 M solution in water). After being stirred for 30 min, the white precipitate was diluted with Et$_2$O (300 mL) and filtered through a Celite pad. The aqueous phase was separated and extracted with Et$_2$O (2 × 100 ml). The combined organic phase was successively washed with saturated aqueous NaHCO$_3$ (200 mL) and brine (200 mL), dried over anhydrous Na$_2$SO$_4$, and concentrated in vacuo. The residue was chromatographed on silica gel (100 g) eluting with hexane/EtOAc (20:1) to give the ketone as a colorless oil (1.17 g, 83%).

Reference: Watanabe, H.; Watanabe, H.; Bando, M.; Kido, M.; Kitahara, T. *Tetrahedron* **1999**, *55*, 9755–9776.

# Index

Acetal, 171, 186
Acetate, 168–169
Acetic anhydride, 61
Acetone, 70, 71
Acetonide, 169–171
Acid chloride, 44, 45
Acyclic diene metathesis polymerization, 160
ADMET. *See* Acyclic diene metathesis polymerization
AIBN, 82, 104, 105
Alane, 86
$AlCl_3$, 139
Aldol condensations, 111–116
    Crimmins' asymmetric aldol, 116
    Evans' aldol, 112–113
    Masamune aldol, 113–114
    Mukaiyama aldol, 114–115
Aldol reactions. *See* aldol condensations
Alkenyllithium, 103
*R*-Alpine borane, 97
*S*-Alpine borane, 97
Alternative solvents (drop-in), 9
Ammonium formate, 108–109
Analogix, 22

Anhydrous solvents, 6–9
*p*-Anisaldehyde stain, 19
Arylsilanols, 128
Aryltrimethoxysilanes, 128
Aspartame, 5
$AsPh_3$, 144
Asymmetric deprotonation, 116
Asymmetric enolate
    alkylations, 122–123
    with oxazolidinone chiral auxillaries, 122
    Myers' alkylation, 122–123

Baeyer–Villiger oxidation, 76, 77
Baker's yeast, 97
Barton deoxygenation, 81–83
Barton–McCombie deoxygenation. *See* Barton deoxygenation
Baylis–Hillman Reaction, 117–118
9-BBN. *See* 9-Borabicyclo[3.3.1]nonane
Benzoin condensation, 118
Benzophenone ketyl radical, 6–7
Benzyl chloride, 171
Benzyl ether, 171–172
(2*R*)-(–)-Benzyloxy-2-methylpropylbromide, 10

Benzyl protection of an amine, 177
$BH_3 \cdot SMe_2$, 86
$BH_3 \cdot THF$, 86, 87, 96
BINAP, 98
($R$)-(+)-BINOL, 130
Biotage, 22
Birch reduction, 107–108
(($2R$-$cis$)-2-[[1-[3,5-Bis(trifluoromethyl)phenyl]ethenyl]oxy]-3-(4-fluorophenyl)-4-benzylmorpholine), 158
Bis(2-bromophenyl)methane, 13
Bis(tricyclohexylphosphine) benzylidene ruthenium (IV) dichloride, 160, 161
Bis-dimethyl ether, 173
Bis-ester, 186
Bis-ketone, 187
Bis(pinacolato)diboron, 146–147
Bis-sulfonamide, 183
$BnPd(PPh_3)_2Cl$, 142
9-Borabicyclo[3.3.1]nonane, 38, 39
BOC-pyrrolidine, 178
Borane, 38, 86
Boron enolates, 111–114
Boron trifluoride etherate, 88, 101, 103
4-Bromoanisole, 146
1-Bromonapthalene, 149
$N$-Bromosuccinimide, 71
Brown asymmetric crotylation, 118–119
Burgess dehydrating reagent, 32
$tert$-Butyldiphenylsilyl (TBDPS) ether, 174–175
$t$-Butyl hydroperoxide, 76
$n$-Butyllithium, 12, 119, 122, 123, 132, 134, 135, 136, 137, 141, 145, 150, 152, 154, 159, 163, 165
$sec$-Butyllithium, 12, 116, 134, 151
$tert$-Butyllithium, 12, 121, 125, 134, 135, 136, 147
$t$-Butyl sulfonamide, 43
$t$-Butyloxycarbonyl (BOC) protection of amine, 178

$(\eta^3\text{-}C_3H_5PdCl)_2$, 128
Camphorsulfonic acid, 170, 176
CAN. See ceric ammonium nitrate
Carbobenzyloxy (Cbz), 180–181
Carbodiimide, 46
Carbon disulfide, 81, 82
Carbon tetrabromide, 28, 150

Carbon tetrachloride, 29
Carbon–carbon double bond forming reactions, 150–166
Carbon–carbon single bond forming reactions, 111–150
Carbonyldiimidazole, 46
Catechol borane, 38
CBS Reduction. See Corey–Bakshi–Shibata reduction
Ceric ammonium nitrate, 172
Cerium chloride ($CeCl_3$), 133, 134
Cerium molybdate stain, 19
 (see also Hanessian stain)
Cerium sulfate stain, 19, 20
Cerium trichloride, 92
Chiral auxillary, 112–113, 116
$meta$-Chloroperbenzoic acid, 66, 72, 78
2-Chloro-5-bromonitrobenzene, 125–126
$N$-Chlorosuccinimide, 62, 63, 66
Chromatography, 17–24
Chromium (II) chloride, 164
Chromium trioxide, 16
Claisen adapter, 157
Claisen condensation, 119, 121
Clemmensen reduction, 102
CM. See Cross-metathesis
Collins oxidation, 57
Column volume, 17, 21
Common crystallization solvents, 25
Cooling baths, 9
Corey–Bakshi–Shibata reduction, 96
Corey–Fuchs reaction, 150–151
Corey–Kim oxidation, 62
Corey–Peterson olefination, 151
$m$-CPBA. See $meta$-Chloroperbenzoic acid
$CrCl_2$, 133
Cross-metathesis, 160
18-Crown-6 ether, 153
Crystallization, 22–24
Crystallization setup, 25
Crystallization solvents, 25
CSA. See Camphorsulfonic acid
Cumene hydroperoxide, 79, 80
Cuprates, 120–121
CV. See Column volume
Cyanocuprates, 120–121
Cyclohexylidenecyclohexane, 156

# INDEX

N-Cyclohexyl-(2-triethylsilylpropylidine) amine, 151
Cyclopropanation, 167

DABCO. See 1,4-Diazabicyclo [2.2.2]octane
Danishefsky's diene, 166
DAPA. See Dipotassium azodicarboxylate
DDQ. See 2,3-Dichloro-5,6-dicyano-*p*-benzoquinone
Dean–Stark trap, 41, 42, 155
Decomposition of ethyl ether, 12
Decomposition of THF, 12
Dehydrate, 32, 33, 51
$\Delta$CV, 21
DEPC. See Diethylcyanophosphonate
Dess–Martin reagent, 15, 16, 63, 64, 65
Diamine, 183
1,4-Diazabicyclo[2.2.2]octane, 117, 118
Diazald. See *N*-Methyl-*N*-nitroso-4-toluenesulfonamide
Diazomethane, preparation, 14–15
DIBAL–H. See Diisobutylaluminum hydride
Dibromoethane, 10
Dibromomethane, 157
2,6-Dibromophenol, 174
2,3-Dichloro-5,6-dicyano-*p*-benzoquinone, 57, 58
Dicyclohexylboron triflate, 113
Dicyclohexylcarbodiimide, 59
Dieckmann condensation, 121
Diels–Alder reaction, 166–167
Diethylazodicarboxylate, 33, 34, 35
Diethylcyanophosphonate, 48, 49
Diethylmalonate, 149
Diethylzinc, 167
3´,4´-Dihydroxyacetophenone, 173
Dihydroxylation, 73
Diimide, 106–107
Diiodomethane, 167
Diisobutylaluminum hydride, 88–90, 93, 103
(+)-*B*-Diisopinocamphenyl-methoxyborane, 119
(*R*,*R*)-Diisopropyltartrate (*E*)-crotyl-boronate, 138–139
β-Diketones, 97
3´,4´-Dimethoxyacetophenone, 173

2,2,-Dimethoxypropane, 170
4-(Dimethylamino)pyridine. See DMAP
Dimethyl dioxirane, 70, 71
Dimethylketal, 185–186
Dimethylmercury, 4
2,2-Dimethyl-1,3-propane diol, 186
2,5-Dimethylpyrrole protection of amine, 178–180
2,5-Dimethylpyrrole, 179
Dimethylsulfoxide based oxidations, 58–63
Dimethyltitanocene, 157–158
Di-*n*-butylboron-triflate, 112, 113
1,2-Diols, 169–170
1,3-Diols, 169–170
1,3-Dioxane, 186–187
1,3-Dioxolane, 187–188
Diphenylphosphoryl azide, 31, 48
Dipotassium azodicarboxylate, 107
1,3-Dithiane, 188–189
DMAP, 47, 49, 169, 178
dppf, 144

EDCI, 46, 47
EDTA, 109
EEDQ, 47
Enamine, 41
Epoxidation, 71, 72, 75, 76
EtAlCl$_2$, 139
Ethyl 2-(trimethylsilyl) propionate, 163–164
Ethylene glycol, 187

Face shields, 3
Finkelstein reaction, 37, 38
Fischer esterification, 50
Flash chromatography, 20–22, 24
Flash chromatography setup, 21
Fleming oxidation, 63
Florisil, 156, 162
9-Fluorenylmethyl carbamate (Fmoc) protection of an amine, 181–182
Fluorous solvents, 8
((2*R-cis*)-3-(4-Fluorophenyl)-4-benzyl-2-morpholinyl 3,5-bis(trifluoromethyl)benzoate), 158
Flur, fluorescent dye, 17
Friedel–Crafts reaction, 123–124
Fukuyama reduction, 90–91
Functional group manipulations, 28–54
Fundamental synthetic organic chemistry techniques, 3–27

1-(3-Furyl)but-3-yn-1-ol, 176
(+)-(E)-1-(3-Furyl)-4-iodopent-
    3-en-1-ol, 169
(+)-(E)-1-(3-Furyl)-4-iodopent-
    3-en-1-yl acetate, 169

Gabriel reaction, 182
Gloves, 4
Goggles, 3
Green solvents, 7–9
Grignard reaction, 124–125
Grignard reagents, preparation, 10, 11
Grignard reagents, titration, 11
Grubbs'catalysts, 160, 161–163

Halogen/metal exchange, 13
Halogenation, 71
Halogen–Lithium exchange, 12, 13,
    135–136, 147
Hanessian stain, 19
    (see also Cerium molybdate stain)
Hartwig–Buchwald aromatic
    amination, 36
Hartwig–Buchwald etherification, 37
Heck coupling, 125–127
Heck reactions of aryl chlorides
    (Fu Modification), 126–127
Henry reaction, 127–128
Hetero–Diels–Alder reaction, 166–167
Hexamethylditin
    ($Me_3SnSnMe_3$), 142, 143
$HgCl_2$, 102
Hiyama cross–coupling
    reaction, 128–129
    standard, 128
    Denmark modification, 129
    Fu modification, 129
Horner–Wadsworth–Emmons reaction
    (HWE), 151–154
    Ando Modification, 153–154
    Masamune–Roush
        modification, 152–153
    standard HWE, 152
    Still–Gennari modification, 153
Hoyveda–Grubbs' Catalyst, 160, 163
HWE. See Horner–Wadsworth–
    Emmons reaction
Hydroboration, 38, 39
α-Hydroxyaldehydes, 97
α-Hydroxyketones, 97, 101
β-Hydroxyketones, 95, 96

syn-β-Hydroxysilanes, 163
Hydrolysis, 52, 53, 54
1-Hydroxy-l,2-benziodoxol-3(1H)-
    one, 15–16
anti-β-Hydroxysilanes, 163
Hydroxylamine hydrochloride, 179
3-Hydroxymethyl-4′-
    methylbiphenyl, 128

$I_2$ or $I_2$ in silica gel, 19
$I_2$ stain, 20
IBX, 15, 65 (see also 1-Hydroxy-l,2-
    benziodoxol-3(1H)-one)
Imidazole, 30
Imine, 42
Ing–Manske procedure, 182
Internal temperature, 119, 157, 158
Internal thermometer, 132, 135, 136
Iodine, 30
4′-Iodoacetophenone, 129
Iodobenzene diacetate, 66
4-Iodo-2-fluoropyridine, 132
Iodoform, 164
Ionic hydrogenation, 103
Ionic liquids, 8
Isobutylchloroformate, 47, 48, 51
Isobutylene, 135
O,O-Isopropylidene acetal, 169–170

Jacobsen–Katsuki reaction, 72
Jeffrey's ligandless conditions, 126
Jones reagent, 16, 17, 68
Julia coupling, 154–155

Kagan asymmetric sulfur
    oxidation, 79
Katsuki–Sharpless epoxidation, 75
Keck stereoselective allylation, 130
Ketal, 41, 188
β-Keto esters, 97
$KHF_2$, 148
$KMnO_4$, 19
Knoevenagel condensation, 155
Kugelrohr distillation, 129, 131
Kumada coupling, 130–131

Laboratory coats, 4
Lachrymators, 171
LAH. See Lithium aluminum hydride
Lawesson's reagent, 43, 44
LDA. See Lithium diisopropylamide

Lindlar's catalyst, 106
Lithium aluminum hydride, 84–86, 91
Lithium borohydride, 87–88
Lithium chloride, 30
Lithium diisopropylamide, 16, 119, 163–164
Lithium tri-*tert*-butoxyaluminohydride [LiAlH(O*t*-Bu)$_3$], 93–94
Lithium/ammonia, 107–108
Lombardo–Takai reagent, 156–157
Lower order cuprates, 120
Luche reduction, 92

Magnesium turnings, 10
Manganese dioxide, 55
Mannich reaction, 35–36
Mannich reaction, vinylogous, 35–36
Martin's sulfurane, 33
Material Safety Data Sheet (MSDS), 5
McMurry coupling, 155–156
Meerwein–Pondorf–Verlag, 95
Menthol, 11
Mercury (II) perchlorate, 188
Mercury poisoning, 4
Mercury salts, 188
Metallations, 134, 136, 137
*ortho*-Metallations, 137
Methanesulfonyl chloride, 29, 30, 31
4-Methoxybenzyl chloride, 108
*para*-Methoxybenzyl (PMB) ether, 172
Methoxymethylene (MOM) ether, 173–174
4-Methoxy-2´-methylbiphenyl, 146
3-Methoxymethylphenyl bromide, 134
*N*-Methyldicyclohexylamine (Cy$_2$NMe), 126, 127
*N*-Methyl-*O*-methyl amides. *See* Weinreb amides
*N*-Methylmorpholine-*N*-oxide, 67, 73
*N*-Methyl-*N*-nitroso-4-toluenesulfonamide, 15
Methyl ether, 173
Methyl orange indicator, 101
Methyl xanthate, 81–82
MgBr$_2$·OEt$_2$, 124, 125
Michael addition, 39
Midland reduction, 97
Miscible solvents, 26
Mitsunobu reaction, 33

Moffatt oxidation, 59
MSDS. *See* Material Safety Data Sheet
MTBE, 103, 104, 105
Mukaiyama aldol, 114–115
Mukaiyama esterification, 48
Multiple carbon–carbon bond forming reactions, 166–167
Myers' asymmetric alkylation, 122–123

Negishi coupling, 131–132
  standard, 132
  Fu modification, 132
NiBr$_2$(PPh$_3$)$_2$, 131
Nickel dichloride, 109
NiCl$_2$, 133
Ninhydrin stain, 19–20
Nitro aldol. *See* Henry reaction.
Nitrone, 78
NMO. *See* *N*-Methylmorpholine-*N*-oxide
NMR. *See* Nuclear magnetic resonance
Norephedrine-derived oxazolidinone, 113, 122
Noyori asymmetric reduction, 98
Nozaki–Hiyama–Kishi Reaction, 133
Nuclear magnetic resonance, 26, 27
Nucleophilic aromatic substitution, 38
Nutrasweet®. *See* aspartame

Olefin metathesis, 160–163
One-dimensional TLC, 17–18
Oppenauer oxidation, 65
Organocerium reagents, 133–134
Organochromium reagents, 133
Organolithium reagents, 12–13, 124, 133, 134–136, 142, 145
Organolithium reagents, preparation, 12, 13
  stability, 12
  titration, 13
Organomagnesium reagents, 124, 130, 131, 132, 133
Organozinc reagents, 13, 14
Osmium tetroxide, 73, 74
Oxalyl chloride, 44, 45, 58
Oxazaborolidine, 96
Oxazolidinone, 112, 113, 121, 122
Oxidations, 55–58
  alcohol to ketone oxidation state, 55–68
  alcohol to acid oxidation state, 68–70

Oxidations, (cont.)
  olefin to diol, 70–76
  aldehyde to acid oxidation state, 76–78
  heteroatom oxidations, 78–80
Oxime, 42
Oxone, 70, 71, 76, 79

P(2-furyl)$_3$, 143
P(o-tol)$_3$, 143, 144
P(t-Bu)$_3$, 144, 146
P$_2$O$_5$, 51, 61
P$_2$S$_5$, 43
P$_4$S$_{10}$, 43
Parikh–Doering oxidation, 59
Pd(dba)$_2$, 129, 143, 144
Pd(dppf)Cl$_2$, 144, 145, 146, 147, 148
Pd(OAc)$_2$, 125, 126, 143, 144, 149
Pd(P(t-Bu)$_3$)$_2$, 126, 127, 132
Pd(PPh$_3$)$_4$, 125, 131, 132, 143, 144, 145, 146, 148, 149
Pd/C, 90–91, 105, 108–109
Pd$_2$(dba)$_3$, 125, 126, 129, 143, 146
PdCl$_2$(MeCN)$_2$, 142
PdCl$_2$(PPh$_3$)$_2$, 142, 144, 145
PdCl$_2$, 125
(2S)-Pentane-1,2-diol, 175
Perchloric acid, 40
Permanganate oxidation, 69
Personal Protection Equipment, 3–5
Petasis reagent, 157–158
Peterson olefination, 151, 163–164
Phenanthroline, 11
L-Phenylalanine methyl ester hydrochloride, 177, 180
1-Phenylnaphthalene, 149
Phosphomolybdic acid, 20
Phosphorous oxychloride, 45
Phthalic acid anhydride, 182
Phthalimides, 182
Phthaloyl protection of an amine, 182
Pinacolborane, 38
PMA. See Phosphomolybdic acid
PMB ether, 172
Potassium 3,5-bis(trifluoromethyl) phenyltrifluoroborate, 149
Potassium ferrocyanide, 73, 74
Potassium hexamethyldisilazide, 153
Potassium organotrifluoroborates, 148–149
Potassium permanganate stain, 20
Potassium phenyltrifluoroborate, 149

Potassium tert-butoxide, 119, 121
PPE. See Personal Protection Equipment
PPTS, 170, 185, 186
Pregna-4,6-diene-3,20-dione, 187
Propane-1,3-dithiol, 188
Protecting groups, 168–189
  for alcohols and phenols, 168–176
  for amines and aniline, 176–185
  for aldehydes and ketones, 185–189
Pseudoephedrine amide, 122–123
Pummerer rearrangement, 66
Pyridinium chlorochromate, 56
Pyridinium dichromate, 56
Pyridinium p-toluenesulfonate, 170, 176
Pyridinium salts, 48

Quaternary thiazolium salts, 118

Raney Nickel, 100–101
Rawal's diene, 166–167
RCM. See Ring-closing metathesis
ROM. See Ring-opening metathesis
Redal–H. See Sodium bis(2-methoxyethoxy(aluminum hydride)
Reductions, 81–110
  alcohols to alkanes, 81–83
  aldehydes, amides, and nitriles to amines, 83–84
  carboxylic acids and derivatives to alcohols, 85–88
  esters and other carboxylic acid derivatives to aldehydes, 88–91
  ketones or aldehydes to alcohols, 91–98
  ketones to alkanes or alkenes, 98–104
  reduction of carbon double bonds, 105–108
  reduction of heteroatoms, 108–110
  reductive dehalogenation, 104–105
Reductive amination, 83
Reformatsky reaction, 137–138
Residual solvent peaks in NMR, 24–27
  in CDCl$_3$, 26
  in $d_6$-DMSO, 27
Retention factor, 17
Reverse phase columns (C18), 22
R$_f$. See Retention factor

Ring-closing metathesis, 160–162
Ring-opening metathesis, 160
Rochelle's salt, 89, 91, 93, 96, 103, 107
Roush crotylation, 138–139

Safety glasses, 3–4
Safety, 3–5
Sakurai reaction, 139
　traditional, 139–140
　Denmark's modification, 140
Samarium iodide, 95
Saponification, 53
Schlenk flask, 132, 146, 156
Schotten–Baumann reaction, 51
Schrock's catalyst, 160, 161
Schwartz hydrozirconation, 140–141
Schwartz's reagent,
　$Cp_2Zr(H)Cl$, 140–141
L-Selectride, 94
Shapiro reaction, 103–104, 141
Sharpless asymmetric
　aminohydroxylation, 74–75
Sharpless asymmetric dihydroxylation, 73–74
Shi epoxidation, 75–76
Sideshields, 3
Silica gel:compound ratio, 21
Simmons–Smith reaction, 167
Single solvent crystallization, 23
$SnCl_4$, 139
Sodium amalgam, 155
Sodium azide, 31
Sodium bis(2-methoxyethoxy-
　(aluminum hydride), 107
Sodium bis-trimethylsilylamide, 122
Sodium borohydride, 87, 88, 91–92
Sodium bromite, 66
Sodium chlorite, 69, 70, 77
Sodium cyanoborohydride, 101
Sodium hypochlorite, 66, 69, 70
Sodium iodide, 31
Sodium potassium tartrate.
　See Rochelle's salt
Sodium triacetoxyborohydride, 83
Sodium/benzophenone, 6
Solvent purification unit, 7–8
Sonogashira coupling, 141–142
(–)-Sparteine, 116–117
Stabilized ylides, 165–166
Staudinger reaction, 109–110
Still's flash chromatography
　technique, 20–22

Stille reaction, 142–143
Stille–Kelly reaction, 143–144
Sulfide, 78, 79
Sulfinimine, 43
Sulfonamide, 183–184
Sulfone, 79
Sulfoxide, 78
Sulfur trioxide·pyridine, 59
Supercritical carbon dioxide, 8
Suzuki coupling, 144–148, 149
Suzuki–Miyaura coupling.
　See Suzuki coupling
Swern oxidation, 58, 62

Takai olefination, 164
TASF [$(Et_2N)_3S^+(Me_3SiF_2)^-$], 128
TBAF. See Tetrabutylammonium
　fluoride
TBDPS ether, 175
TCCA, 51, 66
Tebbe reagent, 157, 159
Teledyne ISCO, 22
TEMPO. See 2,2,6,6-
　Tetramethylpiperidine 1-oxyl
Tetrabutylammonium
　fluoride, 128, 129, 142
Tetrahydropyran, 176
Tetramethylammonium
　triacetoxyborane, 95–96
Tetra-n-butylammonium
　bromide, 126
2,2,6,6-Tetramethylpiperidine
　1-oxyl, 66, 67, 69
Tetrapropylammonium
　perruthenate, 67–68
Thexyl borane, 38
Thin layer chromatography, 17–20, 120,
　124, 126, 137, 139, 161
Thioimidazoyl carbamate, 81, 82
Thioketone, 43
Thionyl chloride, 45
THP. See Tetrahydropyran
Ti(Oi-Pr)$_4$, 130
TiCl$_3$(DME)$_{1.5}$, 155–156
TiCl$_4$, 116, 139, 145
Titanocene dichloride, 157–158
Titanocene–methylidine reagent, 159
TLC. see Thin layer chromatography
TLC stains, 17–20
　p-anisaldehyde stain, 19
　cerium sulfate stain, 19

TLC stains, (*cont.*)
  Hanessian stain, 19
    (*see also* Cerium molybdate stain)
  $I_2$ or $I_2$ in silica gel, 19
  $KMnO_4$, 19
  ninhydrin stain, 19–20
  phosphomolybdic acid, 20
    (*see also* PMA)
  vanillin stain, 20
TMEDA, 116, 141, 156
$TMSCHN_2$. *See*
  Trimethylsilyldiazomethane
TPAP. *See* Tetrapropylammonium
  perruthenate
*p*-Tolualdehyde, 155
*p*-Toluenesulfonic acid
  monohydrate, 103–104
*p*-Toluenesulfonylhydrazide
  ($TsNHNH_2$), 101, 103, 104
*o*-Tolylboronic acid, 146
*o*-Tolylmagnesium chloride, 132
Tosyl chloride, 51
Tosylhydrazine, 101–102, 144
Tosylhydrazones, 101, 103–104
1,1,1-Triacetoxy-1,1-dihydro-1,2-
  benziodoxol-3(1*H*)-one, 15–16
Trialkylsilanes, 81,105
Tribenzoyl chloride, 49
Tributyltin hydride, 81, 82, 104, 105
2,4,6-Trichloro-[1,3,5]-triazine, 61
Triethylborane, 104, 105
Triethylsilane, 90–91, 103
3-Triethylsilyloxy-1-
  iodopropene, 151
Trifluoroacetic anhydride, 60
Trifluoroacetyl amides, 184
Trifluoroacetyl protection of
  an amine, 184–185
Trimethylphosphonate, 152

$N^1,N^1,2$-Trimethylpropane-1,
  3-diamine, 182
Trimethylsilyl diazomethane, 14, 50
Triphenylphosphine, 28, 29, 30, 33, 34, 35,
  38, 109, 110, 126, 149, 150, 159
Triphenylphosphine oxide, 109
Triphenylphosphonium methyl
  bromide, 159, 165
Tsuji–Trost reaction, 149–150
Two-dimensional TLC, 18–19

Unstabilized ylides, 159, 165
$\alpha,\beta$-Unsaturated ketones, 39

(*R*)-Valinol, 171
Valinol–derived oxazolidinone, 122
Vanadyl acetoacetate
  ($VO(acac)_2$), 76
Vanillin stain, 20
Vinylmagnesium bromide, 131
Vinylsilanols, 128

Wacker oxidation, 68
Weinreb amides, 89
Wittig reaction, 159–160, 164–166
Wolff–Kishner reduction, 98–100
Wurtz coupling, 135, 145

Yamada coupling, 48, 49
Yamaguchi esterification, 49
$Yb(OTf)_3$, 166
Ynone, 186

Zimmerman–Traxler
  transition state, 112
Zinc amalgam, 102
Zinc borohydride, 92–93
Zinc reagent, 13–14
Zinc–Copper couple, 156